THE BEST OF PERMACULTURE

A COLLECTION

THE BEST OF

PERMACULTURE

A COLLECTION

Max O Lindegger / Robert Tap

Permaculture — ecologically conscious design. An integration of diverse elements in whole neverending life cycles, each element taking account of the other.

Distributed By:
Permaculture Resources
56 Farmersville Road
Califon, NJ 07830

SOURCES & ACKNOWLEDGEMENTS

SOURCES

The Best of Permaculture is a collection of works that have appeared in Newsletters published by various Permaculture Groups and the International Permaculture Journal.

More information and addresses of groups nearest you can be obtained by writing to the following sources – please include a self addressed envelope.

Australia
Permaculture Nambour Inc.
P.O. Box 650, Nambour, Qld. 4560

Journal of the International Permaculture Association
c/- 113 Enmore Road, Enmore 2042

Permaculture Services Ltd.
P.O. Box 529, Maleny 4552

New Zealand
Permaculture New Zealand
P.O. Box 37030, Parnell

USA
Permaculture Activists, (Peter Bane)
P.O. Box 3630, Kailua-Kona, HI 96745, USA

Europe
Permaculture Institute
Ginsterweg 4-5, 3074 Steierberg, Germany

GAIA Lista
Vidranke, Penne, 4560 Vanse, Norway

Tony Andersen
Sonderboulevard 53.2, 1720 Copenhagen, Denmark

United Kingdom
Andy Langford
The Bakery, Somerton Road, North Aston, Oxford OX
544X, England

ACKNOWLEDGEMENTS

As editors our first thanks must go to the many authors who have allowed us to reprint articles which were orginally intended for only a small readership. Many took time to update the information.

Our appreciation to Terry White for supplying material which appeared in the Permaculture Journal and for his practical advice and moral support.

A thank you to Julie Jones and Trudi Lindegger who typed the manuscript, Len Blanchard, the proof-readers John Skelton, Paul Sykes and TrudyLong, John Reid for the cover artwork, Vergons Tourist Office (Saint Andre-des-Alpes, France), Cecile Barral for French translations and to all who helped this book become a reality.

Also, we thank our Ethical Investors Barry Goodman and the Andersons.

TREE TAX

Each copy of this book carries a Tree Tax of 50¢. The Tree Tax will be administered by Permaculture Nambour Inc. and will be available for tree planting schemes and research in the fields of Forestry and Agroforestry.

Regular reports on the Tree Tax and its use will appear in the International Permaculture Journal and the Permaculture Nambour Newsletter.

FOREWORD

The word, Permaculture, was coined by Bill Mollison and David Holmgren, authors of "Permaculture I – A Perennial Agricultural System for Human Settlements" (Corgi, 1978). The concept of Permaculture describes a process that is essentially one of perennial and sustainable agriculture (a permanent agriculture) and everything this entails when the concept is applied to the world we live in.

Being multidisciplinary in nature there are many fields of interest and avenues of influence for the designed system. This is made very obvious by the scope of articles we have collected.

While some articles may be considered to be inspirational – the Elzeard Bouffier story, "The Man who Planted Hope" is one that we know has touched many people – there are many of practical value offering the type of lateral thinking and understanding of whole systems that conventional agricultural texts lack.

We offer here an insight into people, places, methods and the spirit that is part of a move toward sustainability – an environment to really live in.

CONTENTS

PRACTICE

* Source Publication
IPJ International Permaculture Journal
PAWA Permaculture Association of Western Australia
PC Eng. Permaculture News England
PC Namb. Permaculture Nambour Newsletter
PA N.Z. Permaculture Association Bulletin N.Z.
PC Syd. Permaculture Sydney Newsletter

THE PARABLE OF THE CHICKEN

By BILL MOLLISON

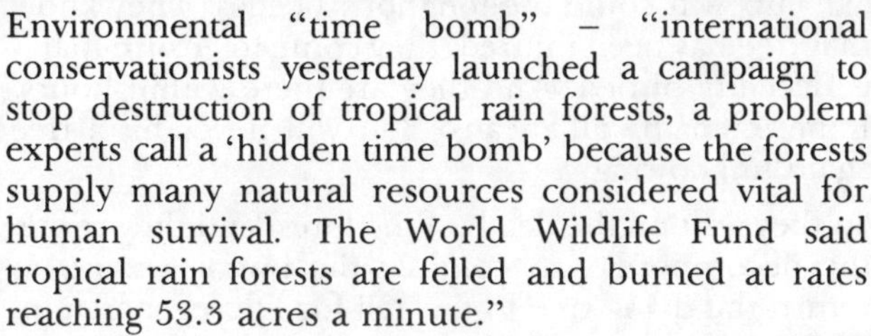

Environmental "time bomb" – "international conservationists yesterday launched a campaign to stop destruction of tropical rain forests, a problem experts call a 'hidden time bomb' because the forests supply many natural resources considered vital for human survival. The World Wildlife Fund said tropical rain forests are felled and burned at rates reaching 53.3 acres a minute."

(S.F. Examiner, Oct. 11, '82)

Conversion of tropical moist forests to cattle pastures ties in the consumer demands of United States citizens to tropical forest ruination to an even greater extent than does the timber situation. In parts of Central America and the Brazilian Amazon, this is the dominant cause of tropical moist forest loss. Between 1950 and 1975 alone, the area of cattle pasture and numbers of cattle in Central America more than doubled. Between 1966 and 1978, an area the size of Maine was converted to 336 cattle ranches in the Brazilian Amazon under the direction of a single development agency. Low land and labour prices make production of grass-fed beef in Latin America about one-quarter as costly as in the U.S. Almost all of these beef imports go to the fast food trade to make hamburgers and other products. The loss of species and genetic diversity in the tropics represents an incalculable loss of opportunities for the entire human race. Many potential pharmaceuticals, foods, bio-control agents, industrial products, and energy sources will never be discovered because plant and animal species will be overcome by the voracious consumerism of humanity even before they are named or screened for possible uses.

Denis Coules, OSA Rainforest Review,
Vol 1 No. 1 Spring 1983

Permaculture is not gardening, it is design. It does not espouse a particular technique whether organic, inorganic, or biodynamic, etc. I personally espouse the organic or natural gardening approach. I talk and write about it, not about pesticides or herbicides. Permaculture is not confined to gardening, or plant growing, it is a design system involving the placement of all the elements of the landscape, of the living system, in the right relationship to each other.

Let me explain. Consider a chicken. We can know some things about the chicken, its particular characteristics. A fancier will note its colour, its qualities for breeding, susceptibility to hawk attack, etc. These are its innate characteristics. It has needs like most of us. Food, a night resting place, elevated perch, modest climatic needs. It has yields: feathers, feather dust, eggs, chicken manure, CO_2 from breathing, and like most animals about 13% of its inputs are turned into methane. If killed, the chicken has other yields too.

There are thus innate characteristics, inputs, and outputs. The inputs are supplied from various sources or other outputs, the outputs go to various destinations, or provide other inputs. The relations between all these elements are studied for our design purposes.

Permaculture does not work with chickens, or glasshouses, or houses, or gardens, it works on making the connections between these elements. A good design would be a complete natural cycle with energy outputs only.

So Permaculture is not gardening, and has nothing to do with technique, in the sense that you know technique, such as how you make compost or kill cabbage moths or why your lemon trees turn yellow. It has a lot to do with exactly where your lemon tree is placed, and its needs as supplied by something else. Permaculture is a skill that says where something goes, so that it functions in relation to other things. Every time you don't do that, you are in trouble. Every input that is not automatically supplied, you must supply. Every output that is not passed to the thing that needs it, must be got rid of. Therefore all undesigned outputs are pollutants, and all pollution is an undesigned output.

All unfulfilled needs are work, and all work is the satisfaction of unfulfilled needs. None of these is necessary, if you have correctly placed every element in relation to its needs and outputs.

Permaculture is the ultimate lazy-man technology. If it is successful nothing needs to be done, and you simply step in the way of the yields, because you need eggs occasionally, and occasionally a chicken.

How much a chicken does will amaze you. It can be used, as a model to partially heat and fuel your home: you can keep control over a large variety of pests: you can double production in your fish pond: you can increase production in your glasshouse: reduce servicing needs of glasshouses, etc. A chicken has a great number of uses. If you neglect these you must do the things yourself. Every time you don't satisfy a need automatically, you must do it yourself.

The question is, is the chicken or the human being the smarter animal in all measurable environmental terms? We will examine that question.

In all broadscale agriculture, which, sadly, is the most destructive influence on the whole face of the earth, the chicken becomes a parable, which, as a representative of a class of animals kept by man, consumes a lot more of the labour of man in agriculture than it returns in necessary food. Man thus works for the chicken. The chicken is enormously smarter than man. Man works extremely hard for the chicken. The chicken works very little for man.

Of all these crops, of every acre of every field of wheat, 70% goes, with the chicken as parable, to the chicken, and 30% to the uses of mankind, not just food uses. So most of agriculture is devoted to the chicken, therefore most tractors, most roads, most rural networks are built to service the chicken. In the total society 35% of all energy goes towards food, so the chicken is a very large consumer of energy in the total society.

Those of us who prevent the chicken operating, are responsible for most of the coal and power station use, and the soon-to-come extinction of the northern hemisphere by acid rain because most of the coal and energy poured into society serves the chicken, or something very like the chicken.

We are about to lose all the forest of Germany. We don't know about Russia but we expect that we are about to lose all of those. This is because we burn so much coal and drive so many cars, that the air has filled with nitric and sulphuric particles. These fall to earth and at first become a fertiliser, and for some years everything grows much better under sulphur and nitrogen. But later there is too much fertiliser and too much of a good thing can be painful. As this acid penetrates into the soil it dissolves something which all soils have, aluminium. Aluminium dissolved in sulphuric acid is a deadly plant poison, also a deadly person poison.

Therefore as the coal and motor vehicle exhausts fall to earth, they start to turn, after a little while, into solutes of selenium, lead, cadmium, and aluminium, all of which are fatal to man, and fatal to plants. The plants then stop rejoicing in the fall of acid rain, and start to suffer, then they are attacked by gipsy moth, tent moth, pinetip dieback, and so on. All these animals sense the death of forests, they are the undertakers of the forest. If you hit a tree once with an axe and wait, by night, above it there will be a swarm of parasitic wasps, and tapping around your axe cuts will come the longhorn beetles. They know that tree has been injured, they come to ensure that it is decently buried. And they are there within hours. Experiment by all means. You will never hit a tree again unknowingly.

So they wait on the death of the forest and they come, the decomposers, to conduct the burial service, to return the dying tree to the soil for life regeneration. We then say the gipsy moth is killing the beech trees, we say the dutch elm disease is killing the dutch elms, we say fire blight is killing our forests, we say pine rust is killing our pines, we say poplar rust is killing our poplars. We don't say we are doing all of it, by driving cars and using energy, and we are blaming the gipsy moth, we are blaming the tent caterpillar, we are blaming the phasmid.

We are in a joint conspiracy not to identify the real criminal. We look at him every morning in the mirror. And we'll all agree to blame the gipsy moth. Now we're free to attack the gipsy moth. We've found the culprit. Now we can go to the forest and spray it with DDT and we can add insult to insult to the forest and the gipsy moth, and ensure that the gipsy moth did indeed kill the forest. And we help it enormously and very quickly to its death. We've just done that to the whole of the northern hemisphere.

6,000 Swedish lakes have no fish. 14,000 lakes dip below pH4 at times. 8,000 lakes have a sharper dip. If people drink water of less than pH4.5 with metals in it, cadmium builds up in kidneys, aluminium combines with protein more strongly in cooking, the aluminium then causing general body deterioration, lead and cadmium affects lungs and the central nervous system.

So it seems that we should quickly join the chicken up to everything and stop wasting all the fossil fuel into the atmosphere. We have no choice. Cars are almost unviable, also coal burning, also nuclear power. A little longer and we have universal death.

Wars and atomic bombs will not kill us, we are killing ourselves. Australia is worse in fallout in many areas than any part of Europe. In Brazil acid rain falls constantly on the Sierra Del Mar. They industrialised. So did we. We signed our own death warrant when we started to dig up the things in the ground. The

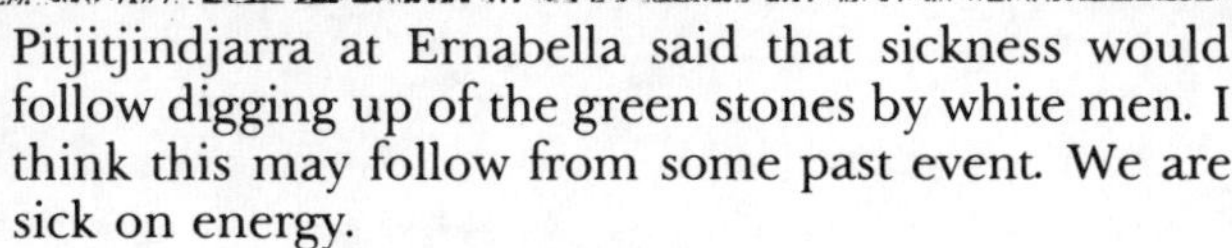

Pitjitjindjarra at Ernabella said that sickness would follow digging up of the green stones by white men. I think this may follow from some past event. We are sick on energy.

The whole of our design efforts must be towards a reduction of our use of fossil energy. Permaculture can't cure anything. It can tell you the way to cure it. To cure it really lies with yourselves.

I started an organic gardening society as an innocent in 1972 because I believed in clean food. I've migrated from that to a study of what is really happening in the world, and realising that being a good gardener can be like being an ostrich with your head in the sand. You will inevitably die in your own good garden if you don't put your head out and see what is happening in the real world. You can't garden under the above conditions. Soil cannot be created under those conditions. Life cannot continue under those conditions. Everything we say about soil is meaningless under those conditions. Therefore for us to continue to live on the earth, stop for a while from just being gardeners, and look at what is happening, and try and stop it. And to a large extent this is why the Permaculture associations are formed, and forming rapidly across Europe, the U.K., throughout all American states, and all Australian states, to tell people what is happening, to help them to design out of it.

Now, while I'm not optimistic, I'm positivistic. I think there are things we have to do. Quickly. I'm not at all optimistic that we'll do them, or that if we do them, we'll succeed. I think we have a very short time in front of us.

An analysis of the cadmium in food sold in Canberra shows 2-3 times above world health limits to avoid permanent lung and kidney damage. This can't be told to the public because they'd stop buying vegetables, and it would require stopping using superphosphate and phosphate rock, and we don't know how to garden without it. We can't tell North American people, by Presidential decree, about the acid rain that's falling on them, because they'd try to vacate north east U.S.A., which would mean chaos that would shatter the economy and shatter the nation. They must be allowed to perish where they are. Real Estate and Banking would collapse. They must be allowed to turn into pickle right where they stand.

We must let Canberra people go on eating their own death warrant. We must let our children absorb lead and become idiots. Politically it is impossible to keep control if anyone is told the truth. The only way we can tell people the truth is to get on our feet and go tell them.

So I get on my feet and I go to Vermont, have the snow reliably analysed, and the same in Germany. Because I'm not the government and I believe people should know what is happening in their lives and should know about the threat of universal death, I'm a political animal, not just an organic gardener.

We've got work to do. In the cities you use far more energy, water, superphosphate, more sprays, than the total of agriculture in the whole country. On lawns we are putting more water, herbicides, pesticides, more agricultural nutrients than the total of Australian agriculture.

Some questions: Do you need any farms at all? Are you prepared to use those resources to grow food? Can we change our soils to be safe soils? This is the only chance, to become home gardeners with a good compost layer on your garden.

Back to the start. We must all become organic gardeners or die. That's the plain unvarnished truth.

I've understated everything I've told you. I'd like you to look at the New Scientist of this year (1983) and read the story on acid rain throughout the northern hemisphere. The original article appeared in *"Der Spiegel Berlin"*, in 3 editions. Now you are up against it, worse than in the bushfires, because this enemy is impalpable. You can't go out and fight something that is yourself.

We can't burn coal or oil, or run atomic power. We can develop Hydro Electric Power, wind energy, solar energy, bio power, and we must be very careful how much fossil fuel we use in that development. We can't run around like blowflies up and down the street, 6 cars going this way this morning and 6 going back this evening, not knowing why they're going in either direction. We can do a modest and essential trip occasionally on alcohol fuel. And we can sail as far as we like and as long as we like. We could even probably balloon great distances, but you get a long way walking, in a long time.

There is a great group called the Society for Growing Australian Plants. I like them and many are friends of mine. They took exception to my following remark: "If you live like an aborigine and garden like a European you'd be completely out of trouble. But when you just insist on living like a European and gardening like an aborigine you're in disastrous trouble. However if you live like an aborigine and eat like an aborigine you're out of trouble. If you can eat as you garden you're out of trouble."

PROFIT OR PERMANENCE?
THE FOOD SUPPLY QUESTION

By STUART HILL

Stuart Hill is an entomologist at McGill University in Canada. In these excerpts from a paper delivered at a Californian conference on small-scale intensive food production, Stuart looks at the nutritional effects of a food system – which is designed to produce food that can be sold – and contrasts it with a food system designed to meet our real need.

Our present "production for profit" food system has evolved from one in which production was for "use". In striving to survive economically, agriculture has had to increase production per area and per farmer. This has led not only to an increased dependence on non-renewable resources, but is has also become a threat to its renewable resource base, and to human health. The farmer is unfortunately in a weak position. He has little control over costs of his outputs and no control over his inputs. Few other sectors of society are so vulnerable. Consequently, it is unlikely that the farmer will be able to correct this situation alone.

It is our view that this state of affairs has already led to a loss of food quality. The ways in which the nutritional quality of plant materials might be decreased are illustrated in Figure 1. This deductive model is based on the concept of Williams (1971), that the body can only suffer from two nutritional problems – lack of certain required nutrients (malnutrition) and the presence of toxins (poisoning).

The model asks how our various food production and handling practices might affect the nutritional and toxic status of the food currently available.

This is particularly important because our nutritional requirements have increased, partly as a result of exposure to the growing number of poisons in the environment (Randolph, 1962; W.H.O. 1972) that require detoxification. Thus, we have a greater need for high-quality food that cannot be satisfied by current agricultural practices. Ironically, the system that should supply us with this food is, instead, contaminating it with poisons and decreasing its nutrient content.

The increase in degenerative diseases in the developed world, and in the less developed areas under the influence of the former, should not come as a surprise. Degeneration is the consequence of genetic predisposition, malnutrition, toxification, (through air, water, and food), lack of exercise, stress, and inadequate relaxation (Figure 2). This makes common sense, yet the dominant approaches being taken to deal with degenerative diseases include the search for causative organisms, the physical or chemical destruction of degenerative tissues (remember that the surgeon reigns supreme within the medical profession), and the masking of the situation with pain killing drugs. This tendency to deal with symptoms rather than with causes, which is equally prevalent in medicine and agriculture, has become the major degenerative disease of science.

Thus, we consider that the prevention of degenerative diseases will require not just medical approaches, but the combined efforts of agriculturalists, nutritionists, geneticists, environmentalist scientists, clinical ecologists, and experts concerned with physical and mental health.

To survive a healthy, contented state, we essentially have to develop a symbiotic relationship with our support environment. This requires that we identify our real needs, and the ability of the environment to satisfy them. Economists frequently distinguish between "real needs" (basic food, shelter and clothing) and "manipulated or non-essential wants" (many of which we strive for to lift us above our fellows). All too often "real needs" are sacrificed for the latter.

The food industry has had to manipulate our eating habits in order to dispose of the surpluses generated by our highly specialised and "efficient" production system. Thus more money is now spent on foods such as corn, wheat, and potatoes in their highly processed and nutritionally inferior forms (e.g. cornflakes, biscuits, and potato chips) than in their more valuable elemental stage (Hall, 1974).

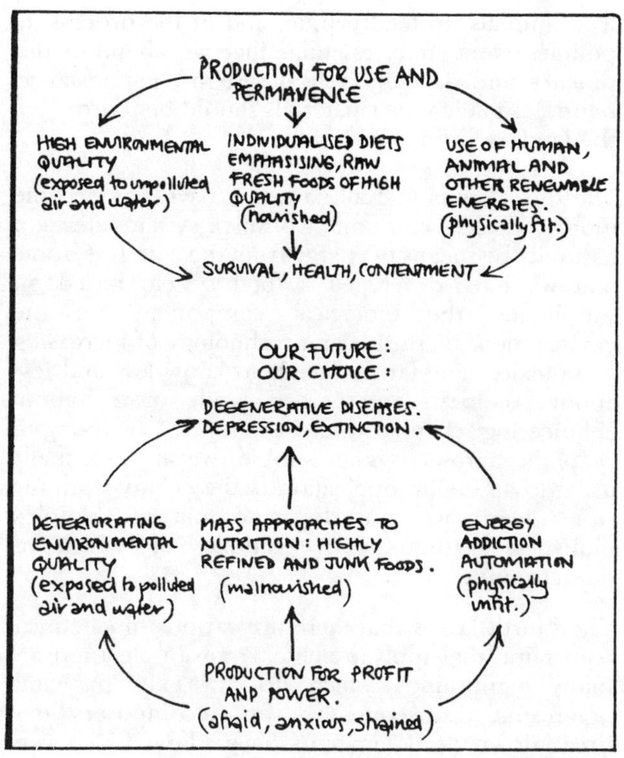

FIG 2.

The processing industry, through the clever use of advertising, has become so successful that it now dominates the food system; and agriculture has been relegated to the position of merely supplying the raw materials. Hence, agriculture now caters largely to "manipulated wants".

Unless we change our myopic view of efficiency, we are misleading ourselves in believing that a more "efficient" agricultural system is the panacea for the food and energy crisis. Changes in policy will result in "real" progress only if such increases in efficiency are concerned with the production and fair distribution of items that we really need.

The first law of nature is that survival for any species, whether it is a plant, animal, or microorganism, is dependent on needs, the availability of what is needed, and on various mortality factors. If we examine our current food systems we find that it contravenes this law at every stage. The implications for policymakers are that they support efforts to distinguish between real needs and manipulated wants and establish a safe food system based on renewable resources.

The second law is that relationships are cyclical. Modern agriculture is charactrized by linear nutrient flows. Thus we produce fertilizer to feed plants, to feed animals, to feed people, and in the process, to pollute rivers. It is essential that we abandon this practice and develop cyclical systems. For instance, natural organic waste materials should be returned to the land as fertilizer.

The third law is that all natural ecosystems become more complex with time. Complex systems develop naturally by means of energy from the sun. It is ironic that we have developed a food system based on simplifying the biological components of the environment by means of a technology of increasing complexity. The farmer needs to know less and less about biology and more and more about engineering, chemistry, and economics. In trying to keep the agro-ecosystem simple, we are essentially using fossil fuel in opposition to the energy from the sun. Clearly we must learn to manage complex biological systems, an example of which is intercropping.

The fourth law is that there are various biochemical constraints that apply to all life. For example, there are many compounds which do not exist in living organisms. Consequently, the decomposers that break down dead organims have adapted to a very restricted diet. Thus if organic compounds are produced that have no counterpart in nature, they will not likely break down biologically. We must establish a life-style that relies only on those organic materials that have a counterpart in nature and ban or severely restrict the production of other organic chemicals.

The laws of nature know no compromise. They are constant, at least within the framework of human history, and the sooner our species becomes aware of these laws and establishes political, social, and economic systems consistent with them, the sooner we will be able to move toward real solutions to our problems.

The only group of producers currently attempting to utilize a significant number of these approaches are the "organic" farmers. Despite the findings of the Center for the Biology of Natural Systems (Lockeretz et.al., 1975, 1976) many agriculturalists still maintain that the people of the world could not be fed by employing these methods. A number of years ago D.O.W. Gussendorf (1979) of Manitoba set out to prove that this was not so. He selected a piece of land declared completely unfit for agriculture. In fact, it was so unfit that he purchased it for one dollar an acre. He established a mixed farm, composted his waste biodynamically (Steiner, 1958) and applied it to his land, while importing no fertilizers, pesticides, or other materials into the system. In 1968 he produced 1000 bushels of potatoes an acre and 50 bushels of top quality wheat. The average for Canada at that time was 168 and 22 bushels, respectively.

It is our opinion that we could learn much by studying the methods of such people as Dr. Gussendorf, for they are doing what many conventional agriculturalists regard as impossible. Their efficient use of energy represents only one feature of the system they are employing.

Currently our food system is designed to produce food that can be sold rather than food that meets real needs. It is imperative that we establish:

1. systems for producing food (and fibre) that can be "permanent" and
2. ways for identifying each individual's nutritional needs.

As both of these are determined by the laws of nature, it is essential that we become aware of those laws.

Many changes will be required if we are to develop a viable alternative food system. Even though energy studies have revealed several weaknesses within the present system, far-reaching political and socio-economic changes will be needed in order to create an environment in which those involved can respond to these findings.

Consequently, scientists are naive if they expect that the required changes will necessarily follow high quality research. One potentially serious weakness within our food system (to which energy studies have not addressed themselves) is that food quality may be compromised at every stage within the food chain. Unless attention is paid to this situation, we predict we will experience further increases in the incidence of degenerative diseases. Indeed, this situation already may be further advanced than we suspect.

We are encouraged, however, by the recent food and nutrition policies implemented by the Norwegian Government. They have established models of consumption based on health and resource considerations and have proceeded to implement policies that will lead to changes in production and eating habits. For example, the price of range fed beef, which contains less saturated fat than feedlot beef, has been reduced below that of the latter. They expect, among other things, that this will help reduce the incidence of heart disease.

It is our hope that other countries will soon examine their food systems and make the necessary changes in policy.

REFERENCES

Blythe, C., 1976
"Eating Your Way Out of Debt and Disease",
New Scientist 70 (939) 278-280.

Grussendorf, O.W. 1969
"Here's the Evidence"   *Land Bulletin* 146:1.

Hall, R.H. 1974
Food For Nought; The Decline in Nutrition, 292pp,
Vintage Books, Random House, New York, U.S.A.

Lokeretz, W.R. Keppler, B. Commoner, M. Gertler,
S. Fast, D. O'Leary, and R. Blobaum. 1974
"A comparison of the production, economic returns,
and energy intensiveness of corn belt farms that do
and do not use organic fertilisers and pesticides".
Report CBNMS-AE-4, 62pp.
Centre For The Biology of Natural Systems,
Washington University, St. Louis, Missouri, U.S.A.

"Organic and conventional crop production in the
corn belt: a comparison of economic performance
and energy use for selected farms".
Report CBNS-AE-7 42pp by the above authors.

Randolph, T.G. 1962.
*Human Ecology and Susceptibility to the Chemical
Environment.* 148pp.
Charles C. Thomas, Springfield, Illinois, U.S.A.

Steiner, R. 1958
Agriculture, 175pp.
Bio-dynamic Agriculture Association, London,
England.

Williams, R.K. 1971.
Nutrition Against Disease: Environmental Prevention.
370pp. Pitman Publishing Corp., New York, U.S.A.

World Health Organisation 1972.
Health Hazards of the Human Environment. 387pp.
WHO, Geneva, Switzerland.

ORGANIC HORTICULTURE AND NUTRITION

By JOHN CALVERT

Organic proponents believe that with the application of organic material to the soil, the soil fauna must decompose it if the nutrients for plant growth are to be released. Soluble chemical fertilisers largely by-pass this process. The plant has more difficulty in regulating its own growth processes because of the by-pass of this important soil function. In contrast, a soil carefully managed and with large organic matter returns, will have better water retention properties, an increased ability to exchange ions, reduced soil erosion, and a greatly enhanced soil microbial population. These facts are well known to most soil scientists.

There are two important aspects to look at when comparing conventional and organic production methods: one concerns the levels of elements considered beneficial to health, the other concerns the levels of health-harmful elements produced.

GREATER AMOUNTS OF BENEFICIAL ELEMENTS

The mineral and vitamin contents of plants are determined by climate and soil (which give rise to considerable variation in the nutritional value) within the limit set by the genetic constitution of varieties. Fertilisation practices have an important effect on the soil, thus on the plants growing in it.

German work over a period of 12 years produced the following results in an experiment which compared two equal rates of fertiliser application. One consisted of organic manures and composts while the other consisted of a conventional high solubility N.P.K. fertiliser (it should be remembered that composts add more than N.P.K. to the soil).

% INCREASE/DECREASE – ORGANICALLY GROWN OVER CONVENTIONAL

yeild (fresh weight)	-24
dry matter	+23

NUTRITIONALLY VALUABLE:

relative protein	+18
vitamin C	+28
total sugar	+19
Minerals:	
Potassium	+18
Calcium	+10
Phosphorous	+13
Iron	+77

DETRIMENTAL/NUTRITIONALLY UNDESIRABLE:

nitrate	-93
free amino acid	-42
sodium	-12

Thus although absolute yield was lower, the vastly increased dry matter, vitamin and mineral constituents meant a food of much higher nutritional value. This work focuses on the important area of maximum yield versus quality. Most fertilisation is aimed at producing the highest yield, but this may not be compatible with a quality product, and as the above shows, there is a need to look deeper than just fresh weight yield.

Similar work done in Sweden in an 18 year comparison between organic and conventional production systems brought out similar quality differences in potatoes – the crop chosen for detailed analysis. While similar yields were found with both treatments, the organically grown produce had higher true protein levels, better cooking qualities, stored better, had more disease resistance and tasted better than conventionally grown produce in the experiment. Thus there can be an enhancement of beneficial elements of quality under organic growing methods.

SMALLER AMOUNTS OF HARMFUL PRODUCTS

A wide range of harmful substances can be found in food, ranging from natural products (Oxalic acid in spinach and solanine in potatoes) to the artificial chemical products of man. Residues of pesticides, antibiotics, hormones, as well as nitrate, free amino acid and sodium levels could be included. Organically grown food should be free from any chemical residues, and as the German work quoted earlier shows, it has reduced amounts of nitrates, sodium and free amino acids.

Heavy soluble nitrogen fertiliser application leads to nitrate in ground water and unduly high levels of nitrate in vegetables. Spinach and other green vegetables have a naturally high nitrate level, but this is dramatically increased by application of soluble nitrogen fertilisers. Outbreaks of methemo-globinemia and circulation disorders in infants, and possible links between high nitrate intake and cancer of the stomach and oesophagus have been shown by several medical researchers. The use of processed manures (composts etc) *and* appropriate crop rotation results in a more regular supply of nitrogen and a reduction in plant nitrate levels. Experiments done by Hardy Vogtmann show these points dramatically.

	DRY MATTER YIELD	NITRATE
Composted manure:		
100 kg Nitrogen/ha	18.2	590
300 kg Nitrogen/ha	19.5	541
Plant compost:		
100 kg Nitrogen/ha	14.4	413
300 kg Nitrogen/ha	18.0	629
NPK(conventional soluble fertiliser)		
100 kg Nitrogen/ha	18.4	1,900
300 kg Nitrogen/ha	18.7	3,587
No fertiliser or compost:	15.3	537

There is already a great deal of awareness on the question of the effects of pesticide residues in relation to nutrition. Pesticide residues tend to concentrate in the outer waxy/oily portions of vegetable/fruit products — on which the major pesticide inputs occur. While refining and peeling may protect one from the residues, such processes also deprive one of the richest area for vitamins, minerals and fibre.

There is little conclusive proof of pesticide residues in food causing such health problems as cancer, birth deformities etc., but as mentioned earlier, these links are extremely intricate biochemically and physiologically, thus very difficult to ever prove. It would certainly be true that there has really been little done to test the longterm safety of the many thousands of chemicals introduced by man, and especially effects of possible interactions among them.

Regular consumers of organically grown foods have lower pesticide levels in their fat tissues than consumers of non-organic foods, as much as 7 times less for someone consuming 80% organic foods, compared with a person consuming less than 20% organically grown food. It would therefore appear that an extra margin of health safety is guaranteed to consumers of organic products.

CONCLUSIONS

There is a great need for more work in the area of long term field comparisons. Such programmes should be comprehensive and look at many facets of production systems. Much of the criticism of biological production methods comes from trials where objectives were short-term and measured only a few parameters.

It would be fair to say that the comprehensive studies carried out so far show a clear tendency for crops from biological production methods to have markedly better quality traits than either unfertilised or conventionally (NPK) fertilised crops. The organic food sources show none of the negative effects of the conventional systems in terms of pesticide residue etc.

That long term trials need to be carried out is clear, but the length of time these take, and the funding they require appear almost insurmountable problems in Australia.

The alternative form of evaluation is more simple and immediate. It consists of looking with a holistic approach at biologically farmed land, and using as criteria positive signs of land health. Thus the research and advisory work can be based around improving these signs of bad health. To use the techniques of biological production as appropriate to improve the ecological sustainability of any given area of land.

REFERENCES

Balfour, E.B.
The Living Soil and the Haughley Experiment
Faber & Faber (1943, 1975)

Harwood, R.R.
"Why Science Wears Blinkers (1983) Research Update"
    New Farm, Feb. 1983, 38-39

Soil Association
Achievements and Prospects (1981). An account of the research programme of the Research Institute of Biological Husbandry at Oberwil, Switzerland.

Schuphan, W.
"Nutritive Value of Crops as Influenced by Organic and Inorganic Fertilizer Treatment (1974)." Results of 12 years experiments with vegetables 1960-1972.
Qual. Plant. Pl. Fols, Hum. Nutr. xxii 4,333-359

Vogtmann H.
"The Quality of Agricultural Produce Originating From Different Systems of Production (1980)"
Soil Association (G.B.)

SOIL CONSERVATION: TIME FOR POSITIVE ACTION

By Max O. Lindegger

The area enclosed by the square on the map below is 10,000 ha. Next time you are taking in the view from Mapleton, take note of the area from Nambour to Palmwoods, across to Buderim and up to Bli Bli.

Well, so what? Sadly, that is the area of land that has been retired from cropping in the Darling Downs (Queensland) because soil erosion has rendered it useless.

Over-zealous land clearing, mainly for grain growing in Piedmonts and Coastal Plains, the black soils of the Alabama, Mississippi and Texas Prairies soon resulted in erosion. The "Dust Bowls" of the 1930's which affected the great plains should never be forgotten.

A National effort since, has been successful in saving much of America's soil and it is unlikely that it will go the way of earlier civilisations.

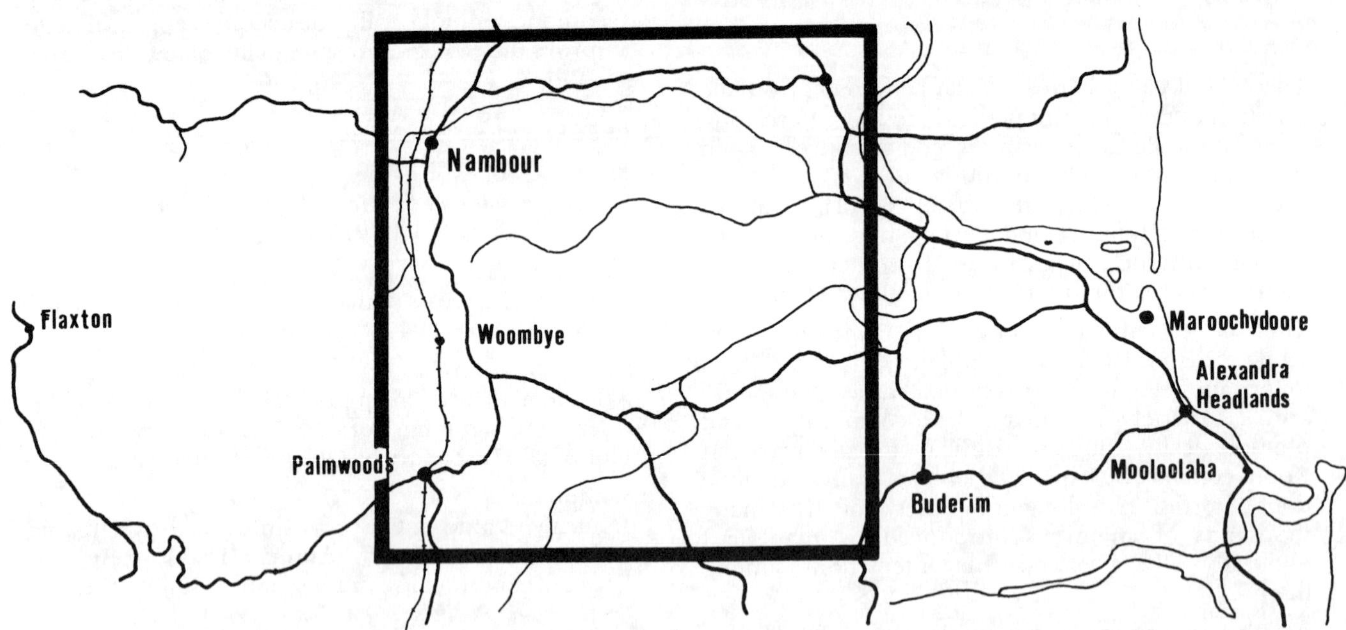

Soil erosion is not new. Over-grazing of valleys of the Tigris and Euphrates allowed the soil to erode and wash down to the plains, choking the irrigation channels in the process. Without water to irrigate the crops on the arid lands, agriculture failed and the great empire of Mesopotamia disappeared.

Soil erosion due to continuous cropping and extreme land clearing turned much of this empire into deserts.

We should also learn from the American experience.

The Australian soils are in a sorry state. More than half of our area in agricultural use requires some treatment for land degradation. Arable land is being lost to wind erosion, water erosion and salination. Australia is losing about 400 ha (about 1000 acres) of good farmland every day.

The three main causes of Australia's land degradation seem to be:-

NEED, GREED, IGNORANCE

NEED arose from the basic requirements to exist on a confined area of land. Settlement patterns and sub-divisional boundaries have never taken into consideration the effects of clearing and ploughing on sloping land. Small crop farmers often have insufficient area to follow sound crop rotation patterns. Graziers on small holdings are frequently compelled to carry stock numbers above an ecologically safe limit in an attempt to earn sufficient income to service their loans. Some finance companies do not take account of the farmer's needs, one example is the policy of high interest on short term loans.

GREED often replaced true pioneering spirit. Development has been taken to mean: to clear and burn. The attitude "If it moves shoot it; if it grows, cut it down" may have resulted in short term profit but at the cost of long term productivity. The white settler in the saddle of a bulldozer is sadly still considered a symbol of progress. Farmers should take responsibility for their own action and allocate a part of their annual budget to soil conservation needs. It is quite immoral to destroy the land to line one's own pockets and then expect the rest of the population (tax via Soil Conservation Services) to foot the bill for their bad farming practices.

IGNORANCE is simply a lack of understanding (common sense?) of what damage certain practices cause to the land itself. Some farmers are driven neither by dire need nor excessive greed, but simply lack the understanding to read changes in the landscape.

In Queensland, more than 60,000 ha of increasingly marginal land are being added annually to the state's grain area, but conservation measures are being planned and implemented for only 50,000 ha. Clearly, not only is the problem not being solved, it is becoming worse. While Queensland's position is somewhat better than the national average, with about 1,600 square kilometers of land in need of treatment, we have a big job ahead. The Soil Conservation Service in Queensland estimates that about $75,000,000 will be needed over the next 30 years for soil conservation. Queensland's actual commitment is not impressive. In 1980/81 we spent only $2.85 million on soil conservation – 10% of what NSW spent. NSW also has about four times our conservation staff. We could gainfully employ another 50-60 soil conservation officers, and with about 75% of Queensland's agricultural land needing conservation work, we need them now.

SUNSHINE COAST: a problem area

The need for soil conservation is especially great on the Sunshine Coast. A number of our

agricultural industries – pineapples, sugar cane and beans – record frightening soil losses and spoilage.

Trials on pineapple plots in Nambour showed soil losses of 148t/ha on land with a slope of 1 in 7 in the first year after planting. Since pineapples cannot tolerate frosts, they are often grown on much steeper slopes and one can only imagine the loss of soil under such circumstances. It is probably fair to observe that a lot of land under pineapples should never have been cleared and if cleared should have been planted to tree crops. Hopefully pineapples will be replaced by crops more suited to this environment.

To say anything against sugar cane in an area to a large degree dependent on sugar is not popular. It is a fact that cane as farmed at the moment cannot be considered as sustainable agriculture. Not only is the industry too dependent on non-renewable energy and fertilizers, the soil losses are too great. On caneland in the Innisfail area, conventional farming practices have resulted in soil losses of up to 300t/ha/year. Green cane harvesting with trash blanket retention, produced losses of less than 10/ha/year. Whilst rainfall at Innisfail is normally more dramatic than in Nambour, soil losses are undoubtedly very considerable in this area.

What we have to realize is that it is not nature but people (farmers, developers, foresters) applying poor practices who are responsible for virtually all land degradation. In arid areas mainly sheep, cattle and other hoofed animals (not drought) are responsible for the destruction of the protective plant cover. It is the removal of tree cover (not heavy rain) which results in slips on the escarpments of Mapleton/Maleny and Dulong/Kureelpa.

Most of the soil lost from the coastal areas is moved by water – mainly by raindrop impact on bare (un-mulched) soil. This results from the removal of the vegetative cover by over-stocking, by clearing along creeks and rivers and by burning and spraying.

SOLUTIONS:

The solution to many aspects of soil degradation is Biological Control. These methods include planting of trees, grass and legumes and broadcast crops which assist in reducing splash erosion and bind the soil against surface erosion. As plant roots penetrate the soil its ability to absorb moisture increases, thereby reducing runoff.

1. TREES

Trees play a very important part in the prevention of erosion. Forest plots on steeper slopes can be very profitable. They offer additional benefits such as increased water retention, moderation of climate, reduction of wind impact, production of nectar and pollen for bees, shelter for wild life, reduction in salting, increased crops and visual appeal. Farms can be self-reliant in timber for building, fencing and firewood. The Forestry Department offers trees at low rates for approved plots and officers can help in the selection of suitable species. Forest trees play an important part in the conservation of soil and water.

2. DIFFERENT CROPS

The introduction of new fruit trees should allow the enterprising farmer/orchardist to utilise sloping land without endangering its long term viability. The transferral of sloping crop-land to permanent tree crops should be encouraged. The change from annual crops to permanent tree crops would aid to minimise land exposed to the elements. In a climate like ours, where heavy storms can be expected at any time of the year, soil should be bare for the shortest time possible if at all. Experiments with permanent legume covers (living mulches) between fruit trees should be undertaken.

3. TRASH RETENTION

Green cane harvesting and trash retention should be introduced immediately. Soil and fertility losses could be much reduced and make the industry more efficient. Suitable machinery is now available in Australia. Burning of organic material can be a massive waste, especially on a large scale. The sugarcane industry is just one example.

4. LOCAL GOVT. ZONING

Local administrations too have a part to play in soil conservation. Local Government bodies have considerable control over subdivision and land use zoning and it is essential that subdivision layouts and zoning standards and conditions should reflect land resource capabilities. Due regard has to be given to the incorporation of soil conservation principles in the design, construction and maintenance of roads and other public works. Drainage works which increase the velocity of water (eg. concrete drains)

Geoff Lawton.

should be avoided. Well designed group title developments should be encouraged. They should incorporate sound environmental standards, preserve agricultural land and allow people to settle in a desirable environment.

Responsible developers should work with natural boundaries such as creeks and ridges, thereby conserving catchment areas. They should, as a matter of course, respect critical vegetation cover and scenic areas. Development adjacent to forests and areas of possible future mineral extraction should be avoided. Protection needs to be given to Fauna and Flora in order to ensure their survival.

It is up to individual landholders to accept responsibility for preserving the nation's most precious resource – the soil. Soil conservation might not produce immediate financial results, but it will avoid longer term disasters. The genuinely concerned farmer should obtain real satisfaction from knowing that the land he hands on will be in at least as good condition (and hopefully better) than when he began to farm it.

The 1948 report of the Soil Conservation Service of the U.S.A. included the following statement:

"Modern soil conservation is not directed merely toward maintenance of the Status Quo. It is dynamic and progressive; it leads to increased and lasting productivity of the land and thereby promotes common welfare". We cannot build a rising standard of living on a falling level of soil fertility. The right to use land carries the responsibility to take care of it for future generations.

The rate at which the world's deserts are growing is 5.4 million ha per year. The Australian contribution to this shocking statistic is plain for everyone to see. Gradual weathering of rock can only create 3.5 cm of soil per 100 years. The process of soil degradation can be stopped and reversed. Modern know-how can speed up the rate of soil creation. It is up to us – *it is up to you.*

European horse-drawn implements. The wrong choice of implement often resulted in severe soil erosion.

67% of the area planted with pineapples in Queensland, is affected by moderate to severe erosion. (QDPI, 1985)

TODD'S PLACE

By TERRY WHITE

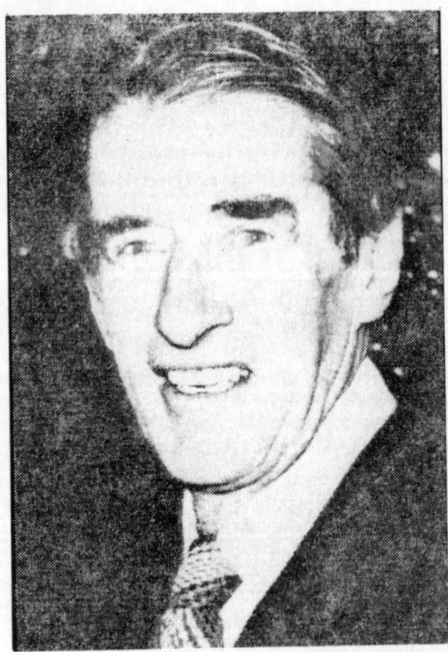

Hector Todd

Queenland's Darling Downs is the richest farming area in Australia, producing two thirds of the nation's maize, sorghum, oilseeds and cotton, but every year about 8 million tonnes of rich top soil (equivalent to about 60 Darling Downs farms) is carried away by erosion.

"In another 100 years Queensland's greatest natural resource will be a desert", says Darling Downs farmer Hector Todd. "We're not farming the Downs, we're mining it". In a 1981 *Bulletin* article (July 21, p.34) Todd spoke out against irresponsible land-use. "The heartbreaking attitude, oh well, it will see me out, is the attitude. It's the next person who will have to worry and then it will be too late."

"If you told these blokes that they were ruining the bloody country they'd knock you down. But in Queensland there are no laws about clearing the mountains and the hills, no laws about clearing right up to a watercourse. We've got plenty in the city, none in the country."

"You can clear a mountain and endanger the valley bottoms from the salt encroachment. You can do what you like and nobody can do anything about it. We're mining the soil far worse than the miners are doing it. Yet the miners are considered ogres".

In 1965 Todd realised that traditional European farming methods were inappropriate on his property. Floods had washed away topsoil and fences and scored out deep gullies across his land.

Todd's solution was to redesign his cropping system on a contour basis. Boundary fences were pulled down and strips carefully laid out at right angles to the direction of flood flows so that water flow was slowed and more could be absorbed into the soil. The farm now has 101 main strips and 30 smaller ones for extra erosion control.

"The big advantage is that when there is heavy rain on the roof at night I can sleep well, knowing that whatever flood comes down the valley we won't lose any soil or rainfall off the farm, as any rain apart from floods is absorbed in the strips".

Now Hector Todd is trying minimum tillage and direct drilling or ploughless farming on his 464 hectare farm.

Awarded the Royal National Association's 'Producer of the Year Award' in 1981 Hector Todd's wheat, sunflower and sorghum yields almost double the shire average and his successful farm management practices have been taken up by neighbours.

GROWING FOR THE FUTURE
LONG-TERM AGRICULTURE IN BRITAIN

By DIGBY M. DODD

This article suggests that our present agricultural and forestry priorities are too narrow for the long-term resource scarcities already visible and that we must begin now to plan for the maximum possible home-grown supply of total renewable resources (materials).

This will require a quantum leap in production and cultivation approaches. Organic husbandry as currently practised will be totally inadequate.

By the turn of the century the world population will have grown by at least one third, one and a half billion more people.

Simultaneously, and partly as a result of the demands created by this population growth, certain minerals and other resources will become scarce. Some argue that there are ample resources, or that need will find new supplies, or that technology will be developed to solve the problem.

But we should stop and realize that an exponential growth in demand at 7% per annum means that each decade alone has been used in all time up to the beginning of that decade. The next decade will again require the total used previously, including all that consumed in the previous decade AND all before. Looked at in this light, most published reserve figures are seen to be fictitious.

Parallel to these impending scarcities we should remember that every time we use energy whether fuel, animal or mineral etc. – we are converting it quickly or slowly to another energy form. Every time a conversion takes place a part of the conversion remains as waste, whether in the form of dust, gas, heat or noise.

Much of this waste cannot readily be recycled or absorbed by nature, so it remains in multiple forms as 'pollution'. Jeremy Rifkin's *Entropy* has described pollution as all the accumulated wastes of all the conversion processes that have taken place. Thus every time we use energy in any form we are adding to the accumulated pollution. When we do use energy it is essential, therefore, that the wastes are able to be absorbed or recycled through natural systems. That specifically does not happen now.

These and other factors suggest that the highest priorities will have to be given to the creation of the maximum possible quantities of renewable resource materials. These can only come from land and marine cultivation. Fermentation and biotechnology are too infinitessimal in scale, and like all new technologies they have dangers not yet assessed. Furthermore they are often based on fossil fuel derivatives as feedstock.

Land acreage is finite and being depleted through poor husbandry, erosion and infrastructure growth. The present emphasis in Britain is on food production. There is a resultant high wastage as only a small part of most crops is actually consumed. Forestry is given even less priority and the remaining land is sparsely used for sheep production or not at all. Even then we produce only 68% of our temperate food needs and only 55% of our total needs, from a total of 80% of our available land. Our forestry supplies only 8% of what we consume. The remaining 45% of our food, and 92% of our forestry, is imported.

However plants, and to lesser extent animals, can provide all the materials and fuels we require in greater or lesser extent including: food, feedstuffs, fibres, oils, chemicals, fertilisers, paints, lubricants, medicines, cosmetics, liquid and solid fuels, plastics, building materials, paper, timber, skins, glues, leathers etc. All are renewable. During the last war the Finns provided almost all of these products and many more from their forests alone and at the end of the war the Americans were surprised at the high standard of living they had maintained. Of course they did not take only one product from each species, but as many as they sensibly could. Plants supply multiple products but present temperate agricultural practice normally harvests for only one product and discards the rest. The inherent waste makes the main product more expensive, as this alone has to pay all the labour, machinery and expensive fuel costs involved.

George Washington Carver, the largely self-educated negro child of Southern United States slaves, revolutionised the Southern American economy and politics by his research into the peanut, the soybean and the sweet potato. At that time these plants were hardly known there or totally uncultivated. Carver saw how the cotton crops were depleting the land, and he sought nitrogen fixing crops which would at the same time give better economic returns. In his laboratory he created over 300 products from the peanut alone, including cheese, milk, butter, coffee, flour, ink, dyes, plastics, wood-stains, soap, linoleum, cosmetics and medicinal oils. From the sweet potato he derived 118 products including flour, vinegar, rubber, molasses, synthetic rubber and postage stamp glue.

By comparison the production from our agricultural and forestry practices is less than tiny. Our emphasis on food only *distracts* from the problem of how to obtain the whole range of products that we need, from our limited land supply. In the future the only logical measure of production will be the total yield of *all* natural products per hectare.

We need a quantum leap in our agro-forestry production. Organic agriculture as currently promoted, though it may be a vital principle in itself, can not create the multiple increase in yields required. We have to look to other ecologically sympathetic methods which instead of the current 2-3 tonnes of mono-crop grain yield per hectare, organic or otherwise, can measure total natural product yields from the same hectareage in *tens of tonnes.*

Trees are nature's method of creating very high volume for restricted land space. Perennial plants now have to take on their true importance. They require lower fertiliser inputs because of their deeper rooting and the recycling of nutrients through leaf-drop. But they have been under-used. Mankind presently uses only some 300 species of plants commercially from the several hundred thousand that exist. 200 perennial plants alone have already been identified as potential food plants appropriate to the British climate. Certainly, some taste much better than others; nevertheless this indicates the genetic wealth available for research, even before we consider any other product derivatives those species might produce, such as we have already seen with the peanut and the soybean.

Whilst many new species are possible they will have to be chosen for their ability to multicrop at different heights so that it should be possible to grow up to five different levels together. Properly chosen varieties will be compatible with each other so as to encourage health, pest control and yield in neighbouring plants (companion planting).

With care and proper management such cultivation can also house – and benefit from – the right density of appropriate animal species. Such integrated agro-forestry is a proper form of ecological cultivation, and should be planned within the existing environment to take full advantage of land contour, micro-climate, water management, aquaculture etc. Indeed, in a well managed system of this kind all the components would contribute to each other in varying ways and would be substantially self-sustaining. In essence this is a systems design approach to supplying our needs, based on a comprehensive knowledge and application of ecology.

Such intensive use of land would, however, require labour-intensive techniques which modern large-scale and heavy fuel consumptive machinery would be inappropriate for. Cultivation in this form might best be managed in the future through communities or amalgamation of farms, with individuals providing certain skills but unable to cover the full range of skills required. The present concept of the individual farmer undertaking all the tasks would not work. This type of system has the considerable attraction, however, that it is very flexible in the scale to which it can be applied. It is appropriate on a large scale, a community scale, or it can be adapted to an individual garden.

Modern energy intensive heavy machinery would in part be replaced by the use of nature's own systems as part of the design. There would, however, be a substantial need for appropriate technology in cultivation, harvesting and for the processing of products and materials derived from the crops. Further appropriate technology will also have to be developed to turn these raw natural materials into products for local manufacture and consumption. All of this will help create local employment. We would need to reverse a longstanding cultural trend in the West, however, in which 'employment' has been equated with 'industry'. By contrast it has long been recognized in the East that cultivation of land is an important part of a person's self development and awareness over and above the agricultural contribution itself.

We are beginning to see a new emphasis already on economic crops as a result of the resurgent interest in biomass for energy. Publicity for Brazil's gasohol and America's conversion of surplus grain crops to alcohol have masked the enormous growth in certain parts of the world for wood as fuel. It is estimated that by 2000 A.D. wood, together with its derivatives such as methanol, is likely to account for 16% of the total world energy consumption. In the U.S.A. wood has already overtaken nuclear power as an energy source and could increase five-fold, to total one fifth of America's energy consumption, within twenty years.

What is particularly interesting for other countries is that Brazil, despite its concentration on monocrops to fuel cars, has through its research and resultant equipment technology and patents created not only internal industries but also substantial export income.

The World Bank's committment during the last two years to invest some $300 million in 37 forestry projects is a further illustration of this developing interest in economic agro-forestry.

A great deal of the knowledge already exists, that is needed to begin the sort of experimental programmes suggested in this article. But a great deal of detailed research is required into species and varieties, into cultivation techniques such as companion planting and compatible species, into products that can be derived from those varieties, and into appropriate technologies for cultivation, harvesting, waste and raw material processing and production. Then it all has to be assembled into ecological systems. As perennial plants alone would have to play a major part in such agro-forestry, the lead times required for species and variety selection, genetic improvement and planting to harvesting maturity will be long.

The pressures of necessity are already well established, however, as we saw previously. These pressures will only continue to expand during the coming decades. It is therefore vital that solutions are planned and implemented immediately as a matter of considerable urgency. It is likely that the development of such knowledge could be substantially self-financing through the creation of export and contract income to other temperate countries, as they will not be long in realizing the urgency of the situation which all of us face. There is also a range of new industries that will be created. Should the U.K. delay in such initiative other countries will undoubtedly establish their own programmes.

Arising from this there are several immediate priorities which need to be pressed for at every level including senior government.

1. The establishment of a temperate products institute, to conduct comprehensive research into the various products that can be derived from individual plant species and their wastes, and how to process these materials.

2. The existing horticultural and forestry research institutions need to direct their often excellent research capabilities into the wider aspects of species and variety selection and improvement, and into cultivation techniques.

3. Appropriate technology needs to be directed, step by step, to address the varying problems of cultivation, harvesting, material processing and production.

If this is done stage by stage, the necessary markets and income will develop as demand increases. Such a programme might best be part of the temperate products institute.

4. There needs to be a programme of co-ordination to amalgamate these components into systems based on ecological practice so that the end result can be far greater than the sum of the parts.

5. Finally, but most urgent and important of all, because the other priorities should flow from it, is the immediate need to persuade the present M.A.F.F. (Ministry of Agriculture, Fisheries and Food) to change its name to M.O.R.E. (Ministry of Renewable Resources). Nothing would direct people's minds quicker to the real and urgent priorities.

NOTE
In this article the term 'economic' is used throughout in the botanic sense meaning 'useful to man', rather than the more common-day meaning of profitable. Crops refer to both plant and animal harvests.

THE THREAT FROM PLANT BREEDERS RIGHTS

(From the Permaculture Sydney Newsletter)
Broadcast on the ABC, Notes on the News Friday December 7th., 1979
Speaker…E. L. Wheelwright, Associate Professor of Economics, University of Sydney.

The *New Internationalist* (no.81) published in Melbourne, concentrates on food production and processing. It makes clear that the pattern of food production and consumption in the world is changing, for it shows how poor countries of the third world are increasingly producing new luxury foods for sale to the richer countries of the world and importing more of basic foodstuffs from them. These luxury foods are fresh and frozen fruits and vegetables, nuts, fruit juices and the like. The basic foodstuffs are dairy products, grains, animal fats and vegetable oils.

The poor countries have little control over the prices at which these foodstuffs are imported or exported, because most of the trade is in the hands of a few giant transnational corporations. Studies by U.N. have shown that this is true for trade in cocoa, bananas, tea, coffee, sugar, grain and pineapples. This information is documented in great detail in a new book by Susan George. She shows how farmers in both rich and poor countries are squeezed by this centralized corporate control of processing, marketing and rural technology.

Even more important than this, in the long run, is the increasingly centralized corporate control of *seeds*. A recent article in *The Washington Post* (25.11.79), entitled *"Seeds of Trouble"* points out that farmers around the world are planting fewer varieties of crops, partly as a result of modern methods of high yield agriculture. For example, in the United States just six varieties of corn account for 71% of the acreage planted with corn and only two varieties for 96% of all peas.

In short, genetic diversity is being decreased significantly. This is bad for two reasons. Firstly, when huge acreages are planted with a single variety, or a few closely related varieties, entire harvests can be wiped out by one disease, or by the temporary resurgence of an insect pest. The Irish potato famine of the 1840's is now thought to be due to the fact that only one variety of potatoes was brought from the Caribbean. The second reason is that new plant characteristics can only be bred from a large genetic pool. The reduction in genetic diversity makes it very difficult, if not impossible to develop new strains tailored to a changing environment, hence the term genetic wipe-out.

The Washington Post also refers to the growing concentration of the world's seed resources in the private holdings of a few transnational corporations, for example a single company United Brands, holds two thirds of the world's banana seeds. A thoroughly documented study of this genetic concentration has been published in Canada, by Pat Mooney.

Mooney concludes that we may be witnessing the gradual extinction of 10,000 years of plant history. All of the major food crops originated in centres of genetic diversity which are almost all in the third world. For example, apples originated in Central Asia, potatoes in the Andes, tea in China, wheat in Ethiopia and peas in Asia Minor.

Although these crops have been transplanted all over the world, their home breeding areas are very important, for the hybrids which derive from them, need occasional infusions of new genetic material if they are to remain hardy and resistant to disease.

A few years of highly organised breeding can not equal over 10,000 years of natural mutation in the original centres of genetic diversity. Because of a breeding bias for yield and uniformity the genetic base of many modern crops is narrowing. Governments in the advanced capitalist countries, under pressure from transnational chemical and pharmaceutical companies, are helping to accelerate this genetic wipe-out.

Hybrids require large amounts of commercial fertilisers, herbicides and pesticides. Large profits can be made from integration, so chemical and pharmaceutical companies take over the seed companies. They can then sell the seeds, and the fertiliser and the pesticides that the crop will require.

There is even more money to be made if the seeds (hybrids) can be patented – just like other technology, this then becomes private property. And unlike ordinary seeds, hybrids must be repurchased every year – the farmer just can not sow part of his previous crop.

This is already happening. Small seed companies are disappearing. In Britain when legislation was passed allowing Plant Breeders Rights, which gives something close to patent control, 84 small seed companies were bought out in one week by a giant food and chemicals corporation. In the U.S.A., there was a similar reaction when the Plant Variety Act was passed. The buyers were the giant chemical and pharmaceutical transnationals, and virtually overnight the seed industry was transformed from a small

enterprise one, to a chemical oligopoly. In Canada this legislation is on the books and many farmers are opposed to it. In Australia, similar legislation has come before Parliament. It is called the Plant Variety Rights Act, and it is not too late to oppose it. Farmers and all human beings should always vote for *diversity against monopoly...*

COMMENTS

Permaculture, as a diverse system, depends very heavily on a reliable source of good quality, non hybrid seeds. Any restriction in the availability of such seeds by legislative or other means is detrimental to a sustainable agricultural system.

Here in Australia, we have so far successfuly resisted the introduction of any form of Plant Variety Rights. Farmers in other countries are less fortunate, and are paying the price. Legislation can be changed. We would like to encourage everybody, everywhere, to work towards a legislation which allows for the free distribution of seeds and all plant material and to become actively involved in the collection of valuable seeds and plants.

THE CORN-FLAKE SOCIETY

ANON.

Studies of the *Tasaday,* the Australian aborigines, and others living as we lived ten thousand years ago, confirm the evidence from the archeological record, namely that they consumed a tremendous variety of foods. The foods of the Australian aboriginal included 29 kinds of roots and 11 kinds of frogs. The *Dayaks* of Sarawak consumed 40 different plant foods, the Tasaday, 50. Indians of the Sonora Desert of the United States, utilize 76 species of desert seed plants including amaranth. Eskimos utilize 27 different plant foods including fermented lichens obtained from the stomach of the caribou.

For an example of just how varied the diet of a 'primitive' culture (one without a supermarket) can be, consider this 'shopping list' compiled by a research scientist (J.B. Birdsell, in the *American Naturalist* 87:171-207, 1953) who studied in detail the diet habits of a Twentieth Century tribe of healthy, long-lived Australian aborigines:

29 kinds of roots
 4 sorts of gum
 2 sorts of kangaroo
 4 kinds of fruit
11 types of frogs
 3 types of turtle
29 kinds of fish
 4 kinds of grubs
 2 species of opossum
   dingoes
 2 species of seals

1 type of whale
8 types of snakes
4 kinds of freshwater shellfish
7 types of iguanas and lizards
  Young of every species of bird and lizards
5 marsupials somewhat smaller than rabbits
9 species of marsupial rats and mice, birds of every kind, including emus and wild turkeys
2 kinds of manna, flowers of several species of Banksia
4 species of nuts, seeds of several species of leguminous plants.

Instead of 40 or 50 crops around the world, as there had been in the mid-nineteenth century, we entered the mid-twentieth with much of the "western" world dependent on no more than 12 crops: *corn, wheat, barley, soybeans, potatoes, rice, millet, sorghum, oats, rye, peas and peanuts* which consitute the bulk of the crops that stand between us and starvation.

Only an agricultural system – ecological agriculture – which encourages diversity, will reduce our dependence on crops we have no control over. Crops often grown in distant lands, subsidised with cheap oil products, overpackaged and transported over large distances are our staples.

Now is the time to learn about new crops, how to grow and prepare them.

Geoff Lawton.

FOODSCAPES, SELF-RELIANCE AND THE LANDSCAPE ARCHITECT

BY MAX O. LINDEGGER

PERMANENT CONSERVATION

The obsessions of the pre-seventies with annuals have been – luckily – mostly overcome. Sadly it has turned into an over use of native plants.   While I use many Australian Natives I avoid out-of-place species.

Locally, I am not always considered to be a conservationist, but my firm belief is that land and open spaces close to our homes and schools and offices should be utilised for production of food. There is no cheaper way to produce food than in your own backyard or on your own balcony. If you don't grow some of your food you must be partially responsible for the destruction of our native forests. It is for our friends who planted only natives in their backyard that the farmers clear land to grow food. It's for them the oil is used to transport the goods to market and trees are cut down to make packing materials. It is quite worrying what well meaning people are responsible for. Landscape planners have a very important role to play in re-educating the public. They do have the opportunity to change trends. Let's explore some of the possibilities.

PARKS. Fruit trees should be grown in most parks. The functions of todays parks, like providing shelter from city life, shade and so on could be as well or better performed by multi-functional trees. Maintenance would be reduced if parks were based on permanent, as opposed to annual plants. Larger parks should have at least parts set aside to become semi-wild. Fallen leaves should be taken off the rubbish list. Let people know that leaves are quite clean and part of the natural function of nutrient cycling. This is fundamentally important in subtropical areas where most of the nutrients are held in suspension and a large biomass is essential. Another point to remember is that a lot of people actually enjoy walking through leaves. Is enjoyment not an important function landscapers should provide? Some parks have shown a tendency to become museums and we should reverse this trend.

Landscapers will have to become more involved in town planning and civil engineering. The straight roads we have come to accept, don't offer much scope. Too many restrictions are placed upon us. We should demand more "streets", narrower roads, bike ways and space for pedestrians. We are in an ideal position to reflect the needs of the general public.

Hopefully, in these uncertain economic times the general public will demand more from landscape architects. It will not be good enough for their work to turn an ugly spot into a beautiful one. It should be

A choice – productive foodscape or purely ornamental.

A stark and unimaginative urban landscape.

Unsightly, unproductive 'hard' engineering.

necessary for it to be productive in one way or another. To fulfil this need it will require a much better understanding and awareness of the useful functions of plants. It is interesting to note that at the outbreak of World War Two in Europe (at a time of crisis) the tulip beds and lawns of Lucerne in Switzerland were planted to potatoes and the flower borders along Lake Zurich were converted to bean production. The landscape architect, Leberecht Migge, always incorporated vegetable gardens and orchards in his design, and so should we.

One new area of development which offers challenging work for landscape designers is the Hamlet under the Multiple Occupancy Act. I consider them ideal as a buffer between the urban sprawl and rural production. A meeting place in many ways. Hamlets are based on what Europeans call "Weiler". A small community which grows gently out of a family unit. We should take more of the past into consideration when we work with Hamlets or similar developments. It should be possible to design a community situation which is largely self sufficient (water, waste disposal, materials, and food), not damaging to the land and environment and offers a quality of life which so many desire, at an affordable price.

For such developments vegetation; hydrology, slopes, soils, and visual appeal should be taken into consideration – as they should for any subdivisional work. We should not be too proud to consult with specialists in other professions to make sure that the final design is of a high standard. Landscape designers should also be allowed to·be involved in overall landscape planning. Only then can we make use of plants as windbreaks, dust and noise reducers, shade, climate modifiers and wild life corridors. Anything less than this is unsatisfactory.

COMMENT

There is light at the end of the tunnel. Local Shires in the south-eastern part of Queensland and also in New South Wales and other states have approved, or are in the process of approving, Group Title-like developments. The guidlines are still in the pioneering stage and well founded suggestions can influence long term discussions. Just as land use policies devised years ago have affected the quality of today's environment, decisions we make today will influence urban and rural life for future generations.

NATURAL ARTISTRY IN FARMING

By DIANE J. LUCAS

P.A. New Zealand Associate member Diane Lucas is a landscape architect working from Geraldine in South Canterbury. In this edited version of her address to the 1981 conference of the NZ Institute of Landscape Architects, she attacks the monotonous lack of local identity in our factory-farmed landscapes. She points out that by following the clues left by natural landforms and native vegetation, productive land uses can be planned which are both visually and psychologically satisfying, and contain the diversity necessary for long term stability.

New Zealand's native flora once visibly expressed the uniqueness and diversity of the land. It was developed in some 80 million years of isolation, from an extremely varied topography, which gave rise to a variety of climates, soils and biota.

But the N.Z. economy is now based on production from exotic plants and animals. Rapid transformation of ever larger tracts of land by replacing our complex, specialised ecosystems with simpler unspecialised ones has been destructive of our national character and of biological health and richness. The proverbial 'Romney-Radiata-Ryegrass' syndrome has taken over, and tends to mask much regional character and local variation.

The rapidness of this change and the harshness of the boundaries show the youth of our agricultural landscapes. Boundaries of cover-change often dominate the landscape. Rarely has the visual significance or biological potential of these 'edges' been positively developed.

Not so long ago, at least a framework of remnant native vegetation – whether forest, shrubland or wetland association – emphasised landform patterns and expressed local soil and climatic conditions. Together with buildings largely constructed of local materials, some significant local character was assured even in developed rural landscapes.

The most basic component of any landscape is the landform pattern. From this, any variety in soil, moisture availability, aspect or microclimate is

(Photo J. Toye)

expressed in vegetation patterns. Ignoring the patterns of the land, using simple monocultural cover which is adaptable to a wide range of conditions (with any unsuitable conditions being modified e.g. drainage), has created a vast factory landscape deficient in both visual and ecological value.

With complex landuse systems the same patterns would be expressed in an intermix of pastures with short and long-term crops, chosen and sited to maximise the production and sustainability of each micro-environment. Complex systems would make best use of the land, and be of considerable visual value in reinforcing the patterns of the land, and expressing regional differences. They would also provide diversity of cover which would create a more healthy and stable ecosystem. Maximum use of local native vegetation helps express identity – as complex associations, or more simply as shelter systems.

PROMOTION OF UNIFORMITY

The advice and encouragement made available to

landowners promotes uniformity in our landscapes. Government incentives for clearing, draining, forestry, shelter, pasture improvements act to promote the creation of more uniform factory landscapes. Pressures for increasing productivity whilst reducing labour requirements, further encourages simple landuse systems.

New Zealand rural landowners have considerable freedom in the use of their land. They may dramatically alter the land and its cover without outside interference. Unfortunately many of the incentives and much of the advice is single purpose, and implementation is often in conflict with other government policies. Practices aimed purely at extending our grass factories are destructive of forest, shrubland, tussock grassland and wet habitats, and thus often destructive of a resource which may be significant in water and soil conservation, climatic moderation, native vegetation retention, wildlife habitat, pest control, bee forage and visual quality. With the destruction of any part, insignificant though it may seem itself, the ecosystem is further simplified and thus becomes more vulnerable to stress.

THE WHITE POST SYNDROME

There may be superficial value, or purely superficial value, in a less diverse or more 'tidy' landscape. The trend for 'tidiness' being equated with 'looking good' is hopefully on the decline. This attitude is often destructive of 'wasteland' which may be valuable habitat, and often promotes superficial 'tarting-up', the white post syndrome. Encouraging greater awareness of basic landscape values can surprisingly quickly shift these same energies to positive landscape development. There is considerable dissatisfaction with the trend towards uniform and tidy landscapes. They do not satisfy many who visit, or many who live and work there. Unfortunately with lack of guidance

to the contrary, the response to this dissatisfaction is often expressed in superficial trimmings, such as garish building and planting. But increasingly there is more positive response such as diversification into various tree crops.

Many are actively seeking ideas and advice to give some variety and satisfaction to their holdings.

Considerable effort is now needed to ensure that the rural landscapes of the future are psychologically satisfying, that production is sustainable, and that they are an expression of the uniqueness and diversity of the country.

Many proposals to improve agricultural landscapes are based on pinpointing areas of unproductive land as amenity areas. Such an approach reinforces the attitude of considering farms as factories. When rural landscapes are considered as ecosystems there should be no unproductive land. Every land area needs to be assessed not only for its most productive use, but also for its ability to fulfil a variety of uses within that land system. All land can be useful, whether for animal grazing, timber or tree cropping, shelter, specialised cropping, bee forage, pest predator habitat, wildlife and local indigenous flora retention.

Maximising landuse variety to best fit the physical characteristics of the land seems a positive basis for the evolution of satisfying rural landscapes.

To divide rural land into productive and amenity areas would seem a limited and superficial approach, particularly as the homogenising effects of current technology cause an ever diminishing amount of 'non-productive' land.

20 Expansive pastoral scenes with a mountain backdrop need not be "the real New Zealand."

An overcleared landscape. (Photos by J. Toye)

Kiwi fruit trellising and artificial wind breaks.

PROMOTING POSITIVE CHANGE

Recent work as a consultant to landowners has indicated that a major inhibitor to the development of more satisfying landscapes may well be a lack of awareness of the negative impacts of many of their activities – "they didn't know anyone cared". Typically, the advice they receive is on production, efficiency and economics; not on alternatives to the negative impact such change may have on their quality of life. Ignorance, more often than wanton destruction, is to blame. Farmers understand the land, its variations, sensitivities and limits. But pressures to 'develop', to mechanise and use artificial controls have allowed and encouraged much land sensitivity to be ignored.

Economic pressures have largely dominated ecological principles in the art of agriculture. Energy limitations, effects of chemical control measures and economics may cause the "agribusiness" of many lowland areas to be reconsidered.

Alternative development of multi-tier mixed farming systems would dramatically alter these landscapes and express greater site-sensitivity.

Obviously, attitudes to the land are important in the approach taken to land development. In some parts of the country, farm houses appear to sit in suburban-style sections totally unrelated to the property. Indications are that discussion and the display of alternatives, may gradually alter this attitude which divorces living and working areas of a property. An attitude of living with and belonging to the whole property would seem to be important for development of pride and concern for the general landscape. Many farmers now welcome and encourage advice on practical means of creating more interest and diversity on their properties.

To improve our landscapes, an awareness needs to be developed of their values, their vulnerability and the importance of variety. Such awareness among those planning, researching, studying, advising, manufacturing for, and those actually working with rural land, is essential.

Specific leglislation and controls are probably not the answer. The freedom farmers now have in their activities carries with it moral responsibility for the land. To remove some of that freedom could be expected to also remove some of that sense of responsibility. This could be of considerable detriment to the land. To counter current advice, advertising and incentives, much encouragement of greater responsibility and sensitivity to the land is desperately needed.

Demonstration farms may also be an effective tool in guiding positive agricultural change. Practical farmers like to SEE their aims in action.

Often there are choices in the land management decisions made daily. These decisions could be influenced.

CONCLUSIONS

Development pressures are mainly geared for the immediate future; not for future generations. But widespread unemployment, economic instability, and market uncertainty are encouraging some rural people to consider options for the more distant future. Some are seeking production diversity to support a greater number and range of livelihoods on a property. Many more are seeking diversity to counteract trends and pressures for uniformity. Such diversity could enrich our countryside.

COMMENT

Economic factors are still playing a major part in the choice in which direction agriculture will go. During a visit to New Zealand (for the designers course with Lea Harrison) a positive trend towards more sustainable agriculture was very visible. A number of farmers from 'conventional' areas attended the course, which to me was a positive sign.

REFERENCES

"Landscape Guidelines for Rural South Canterbury"
$6.00 plus postage from Diane Lucas.
73 Connolly St, Geraldine

Mollison B., Holmgren D.
Permaculture One: Perennial agriculture for human settlement.
Corgi, 1978

Mollison B.,
Permaculture Two: Practical design for town and country in permanent agriculture.
Tagari, 1979

Douglas J.S., Hart R.
Forest Farming
Watkins, 1975

Douglas J.S.
Alternative Foods
Pelham Books, 1976

Smith R.
Tree Crops
Devin Adair, 1953

Fukuoka M.
The One Straw Revolution: An introduction to natural farming.
Rodale Press, 1978

REAFFORESTATION AND AGROFORESTRY IN EAST AFRICA

By STEVE PAGE.

I've just returned from East Africa, an area where, due to land owernership being a birth-right of a large sector of the community, large monocultures are the exception rather than the rule. While many large plantations do exist, especially in the tea, coffee, sisal and sugar industries, the large population density of the original inhabitants has enabled them to either retain, or regain, their small farms and gardens, known as shambas. Many villagers with only a few acres provide sugar cane, cotton, pyrethrum, coffee, tea, and cashews to co-operative or private processing and marketing boards.

With the availability of more detailed research into agriculture, forestry and agroforestry, the opportunity exists for much of the urgent, and proposed, development to be based on sound ecological principles of land use.

The work of the various United Nations agencies, as well as many private aid organizations, has helped to bring the importance of agroforestry to the ears of governments.

The UN Environment Programme (UNEP) has given priority to the problems of desertification in its 1982 10th anniversary activities, with a global reforestation campaign entitled 'For Every Child a Tree'. To ensure that the next generation of adults will be conscious of the environment and the complex link between all living things, UNEP aim at creating an understanding in children of the essential value of trees to humanity. A number of countries have pledged support. Amongst others, Denmark, Czechoslovakia, India and Canada will provide seedlings and technical assistance.

Despite the recent calls by our Prime Minister to the international community to increase its support to the poorer nations, everywhere I travelled the same countries were prominent with government funded aid programmes, Sweden, Norway, Canada.

In Kenya, several organizations, among them the Forestry Department of the University of Nairobi, International Council for Research into Agroforestry (ICRAF), and International Development Research Council (IDRC), are continuing research to determine suitable species for the various climatic zones and different land uses.

Photo, Mike Crooke

At Oloseos in Southern Kenya I visited a Maosai Rural Development Centre run by a committee of local people with some external funding by World Vision. Situated in Maosai grazing land, originally utilized by nomadic pastoralists, the area has come under increasing ecological pressure as people take to farming, settle, build wooden houses and draw upon the scarce local timber for firewood. Erosion and a dropping water table are considered the main concerns by Range Management specialists from the University. The Oloseos Centre's aim is to, by example, develop an awareness amongst the local people of the benefits and ease with which trees can be grown around their homes, displaying improved cropping techniques and developing a health centre complex. From a distance the centre stands out as a green oasis in an arid landscape as do several nearby settlements, proof of the centre's successful impact. Problems arise however, with those still living a semi-nomadic lifestyle as their many attas (family compounds), as well as any young trees, are often deserted as they seek seasonal changes.

A key force behind the current calls for tree planting by various governments in Africa is the economic necessity of firewood. With supplies rapidly dwindling, dependent on wood or charcoal for cooking and water purification. In Nairobi while awaiting the bus, I would see women carting load

26

A pollarded Neem tree. Firewood production with conservation of tree cover.

A deforested hillside.
Photos, Erik van der Werf.

after load of firewood on their backs from a timbered gully, depositing them at the bus stop. When 4 or 5 of these 4 foot long by 2 foot diameter bundles of sticks were assembled they would be loaded onto the roof of the next bus and taken out of town to subsequently be converted into charcoal and traded at the marketplace. When people start coming into town to collect firewood there must surely be a grave shortage.

According to Fred Owino of the Forestry Department, ''the stage is ripe for Kenya to effectively combine agriculture and forestry. The solution to the rapid consumption of forests for agriculture is to evolve a land production potential per unit area of arable land''.

In Kenya the President has joined hands with villagers to construct erosion prevention earthworks and in tree planting.

It is imperative that tree planting and soil erosion control are seen as political issues in these countries, as it is difficult for villagers to act on long term requirements when day to day problems of survival consume much of their energy.

FARMERS OF FORTY CENTURIES

BOOK REVIEW

Farmers of Forty Centuries or Permanent Agriculture in China, Korea and Japan by F.H. King. (441 pp, $12.50, published 1911 originally, undated reprint by Rodale using the same plates.)

I had heard about this book, and it is true – this book is a classic. The information it contains is not the important bit (who wants to work like a Chinese peasant of 1907?) but rather it contains a description of a sustainable agriculture, something we very much need.

King was aware of the dependence of U.S. agriculture on imported fertilisers, machines and abundant land, and he was also aware that it was destroying the soil, "When we reflect upon the depleted fertility of our own older farm lands, comparatively few of which have seen a century's service...". What amazed King, was the intensive farming on soils which had been farmed continuously for as much as four thousand years. He comments on the intensity of farming and its permanence time and time again. "... in the Shantung province we talked with a farmer having 12 in his family and who kept one donkey, one cow, both exclusively labouring animals, and two pigs on 2.5 acres of cultivated land where he grew wheat, millet, sweet potatoes and beans. Here is a density of population equal to 3,072 people, 256 donkeys, 256 cattle and 512 swine per square mile."

Much of the book deals with how this was achieved. Basically the soil was fertilised by fanatical recycling of organic matter, the use of legumes and the importation of organic matter where possible. King describes farmers competing with ornate toilets so that travellers might add their faeces to the farm's fertility. Canal mud from the hills was religiously removed and applied to the fields at rates as high as 70 tons per acre (by hand). A contractor paid the cities for the right to remove sewerage.

Another method King found interesting was the overlapping of crops in time and space. "...The crop was cucumbers in groups of two rows thirty inches apart and twenty-four inches between the groups. The plants were eight to ten inches apart in the row. He had just marketed the last of a crop of greens which occupied the space between the rows of cucumbers... the vines were beginning to run, so not a minute had been lost in the change of crop. On the contrary this man had added a month to his growing season by over-lapping his crops...with ingenuity and much labour he had made his half acre for cucumbers equivalent to more than two. ...Four acres of cucumbers handled by American field methods would not yield more than this man's one, and he

grows besides two other crops the same season. The difference is not so much in activity of muscle as it is in alertness and efficiency of the grey matter of the brain." This is the principle of diversity and complexity in a basic form, and probably useful in pest control as well.

The canals of China also impressed King. He describes how they serve to distribute water out over the land. King estimated there were 200,000 miles of these irrigation channels in China.

King was obviously inspired by what he saw, however he also carefully points out both sides of the story. He correctly observed that the areas with fantastic population densities "...occupy exceptionally favourable geographical positions so far as these influence agricultural production." This is because the rainfall in these areas is much higher and better spread seasonally than the U.S.A. which is roughly 10 degrees further north. It is this factor primarily which allows up to four crops on one piece of ground in one year. The soils of these countries are also inherently very good. King often observed that many of the agricultural practices were dependent on a staggering input of human labour, which would never be possible in the U.S.A.. He also mentions that the most fertile areas are the river plains, and he touches on the inevitable cost of this – one million dead in a single flood.

Another interesting observation made by King was that there were hardly any flies in the course of their travels.

Perhaps what Kings saw can be summarised in his own words, *"Living as we are in the morning of a century of transition from isolated to cosmopolitan national life when profound readjustments, industrial, educational and social, must result . . . It is high time for each nation to study the others and by mutual agreement and co-operative effort, the results of such studies should become available to all concerned, made so in the spirit that each should become coordinate and mutually helpful component factors in the world's progress."*

COMMENT

We include this review as much for King's profound statement above as for its highlighting of the fact that concepts empolyed in Permaculture have been evolved over centuries and in many lands, both by our efforts for survival and by natural processes. Throughout history there have been examples to learn from. Agriculture is no exception.

THE WORK OF MASANOBU FUKUOKA

By ZEN

Masanobu Fukuoka

"We have come to the point at which there is no other way than to bring about a movement not to bring anything about." Masanobu Fukuoka – The One-Straw Revolution.

Fukuoka sensei has been my teacher, spiritual father and guide for nearly 11 years, ever since I left the Zen Temple on Honshu Island for a walking pilgrimage around his island. I heard of him from Larry Korn, later to become the editor of the now famous book *"The One-Straw Revolution"*. Larry visited our Zen monastry in the Kyoto mountains and told me about this enlightened man who owns a mountain and who, with his strange farming methods and philosophy behind it, kept the establishment "up in arms" against him. Although the latter sounds sensational for the media, the news of him being a master without lineage must have made me restless enough to have our head priest send me on the before mentioned pilgrimage.

When I arrived 54 temples and 2 months later, I found sensei with 8 young Japanese helpers and 6 foreigners digging trenches into the side of a shallow creek. They were not ordinary irrigation trenches but of the First World War kind! 4 feet deep and spaced 3 feet parallel in rows of 5 and 30 feet long. I was later told by the foreigners that this was his way of soil-aeration for future fruit tree planting.

On the night of my arrival, in the warm atmosphere of candlelight and the kitchen's open central wood fire, I accepted sensei's invitation to participate in his work – despite the grim prospect of becoming a trench digger.

Why did I do it?

Many a visitor upon first meeting sensei have a profound experience of being accepted and loved; and like a child lost in a crowd many find it difficult to leave his presence.

Many times I had to leave, either for undergraduate studies, research at Kyoto University, making money as a freelance artist or simply for visa renewal to Korea. But, for the past two years I was able to stay uninterrupted while working as a teacher in the English Department of a nearby University.

As his secretary I translated his overseas mail and the invitations from Australians made him consider a visit as he had been to the U.S. and Europe before. He will be 71 this year, a reason his family gives for trying to keep him from travelling far; and there is also the Ministry of Agriculture who for years rejected his farming methods, but recently sent out officials to check up on what his crops are doing. His rice for example yields more bushels per acre with bigger leaves, stronger roots and stems and 140 seeds instead of 90 seeds. They were so keen to find out his method that they offered him a deal: for telling only them his ways of farming and keeping especially the foreigners away (because of U.S. – Japan trade frictions), he was to become a living National Treasure of Japan.

Now, what are his methods? I will have to look back 37 years when he was a young Chemistry graduate working for the Agriculture Department at a lab in the next prefecture. The Americans had just won the war and were pushing successfully farm technology, pesticides, fertilizers, chemical sprays and poisons. The conquerors found a ready market in the confusion of post-war Japan: but not in Masanobu Fukuoka who became increasingly disillusioned. It was also a time of his spiritual quest with enlightening experiences culminating in the final decision to end his career with the department.

That is not an easy decision in Japan where life employment in one institution is the rule and in the world's most group-oriented society where a supreme value is placed on conformity. Fukuoka's step out was decisive. The Japanese word for 'leaving employment' is 'cutting one's throat'.

Back at his hometown he acquired 1/8 of an acre of orchard on a hill with a hut, a dog and 5 chickens. There he lived his philosophy. Within the first 10 years he wrote 4 books and became the first Japanese to oppose the new farming methods. He developed rice that, like 200 years earlier, did not need digging, planting of seedlings, fertilizers, chemical sprays and most unusual of all, flooding during the Summer months.

Those of you who read The One-Straw Revolution may remember reading about his dry-rice growing on slopes; that is incorrect. It is true that Larry had seen some rice growing on the hill during his writing, but that was just one of sensei's little experiments which are so common to his farming. His rice needs a soggy soil during the rainy season which can only be provided on level plots with a drainage system.

He inherited the rice fields from his father half a mile from his mountain which had attracted willing young helpers who had read his books. The 'One-Straw Revolution Farm' had also grown considerably in size. He bought many surrounding plots from willing neighbours who were disappointed with the low yield of fruit because of poor soil conditions. Trees would only grow to a certain height and were prone to diseases. I am told that after he bought a plot he would pull out the weeds and spread clover seeds and a variety of herb and vegetable seeds. Whatever came up would reseed year after year and naturally choose the spot most suitable.

He knew the well-being of every tree and plant by observing the activity of the insects surrounding it. And here it is interesting to note his philosophy which observers have placed in line with the Chinese-Taoist philosopher Lao-tse and the humanitarian Albert Schweitzer. Every visitor to the farm is being enlightened of man's illusory existence.

The discriminating mind, creating opportunities, for example of good and bad, loses the understanding of nature. Only people, he says, distinguish good and bad insects; but nature 'just is' and insects attack the plant either because it's weak or it doesn't belong to that spot in the first place. But there was one time, when he told us to actually manipulate the insect world, that was when a worm attacked the trunks of fruit trees and we had to remove them with a special wire and give them to the chickens. It took us weeks, but saved spraying like the neighbours. Something like 400 kg per rice acre-season. Fukuoka's rice is, therefore, not purely organic because of drift from the neighbours who lately kept spraying well into September, close to harvest. They claimed they were overrun by Fukuoka's "bad insects" – as if his happy insects would want to live in the poisoned soil and rice of the neighbours!

For those of you who haven't read The One-Straw Revolution, I may just briefly mention that whereas the common method is to plant young rice seedlings in April, harvest in September, then till the soil and plant vegies (sometimes in greenhouses) in Winter and till again in Spring for the rice; sensei's field is never tilled; growing rice in Summer and wheat in Winter. The seeds of one crop, being spread, depending on weather, just before harvesting the other crop with perfect six months timing. In those 35 years he has worked out a very delicate system for his rice, and while he was away in Europe the helpers made a mistake in timing. It took us 2 years to re-establish the yield after 2 cycles of failures. Sensei even flooded his paddies for remedy. That is another one of his "secrets". He tries and tests new methods all the time. His experiments with Fungi have given him more fame than his books on philosophy. He proved the Agriculture Department wrong by explaining the nationwide death of the pine-tree to have been caused by a fungus spread by airborne spores from imported Russian logs instead of being caused by a pine-mite. The mites attached to the pine after the root-rot from fungus weakened it. They are still spraying monthly with helicopters all over Japan.

I would now like to share with you some universal truths that I'm sure sensei would suggest for your situation here.

One is to be alive with all plant/insect/animal life at your organic farm. Work with your needs and have little machinery for just your size of farm.

If you have land, and you are not growing your own food, you support the use of poisons and lose valuable experience in organic farming methods, apart from the other environmental issues involved. Fukuoka's revolution, like permaculture assures life on earth for your children and generations to come and gives purpose to existence.

As Lao-tse says: "The purpose of life is to know the purpose of life".

The garden of "The balance of mind and intuition". Fukuoka

After a hard day's work in the fields and orchards sensei would sit, just like 35 years ago, peacefully with a cup of tea at sunset in his hut and radiate happiness. Everyone is touched by a feeling that he "knows".

The present situation at Fukuoka's ONE STRAW REVOLUTION FARM' in Japan.

Max Lindegger of Nambour, Queensland asked me to clarify reports of political and legal intervention in Fukuoka's 'revolution'.

Due to his age (72) and social, cultural and kinship obligations towards his family (wife, daughter, two sons, grandchildren), relatives, neighbors, friends and the local farm community, Mr. Fukuoka accepted certain restrictions placed by the Ministry Of Agriculture. These include the restricted use of patented seeds and farming techniques. There are no more guided tours and facilities for visitors have been closed.

The first draft in English of his philosophy in 3 volumes was presented for approval to a Tokyo publishing company.

ELZEARD BOUFFIER: *The Man Who Planted Hope*

By JEAN GIONO

For a human character to reveal truly exceptional qualities, one must have the good fortune to be able to observe its performance over many years. If this performance is devoid of all egoism, if its guiding motive is unparalleled generosity, if it is absolutely certain that there is no thought of recompense and that, in addition, it has its visible mark upon the earth, then there can be no mistake.

About forty years ago I was taking a long trip on foot over mountain heights quite unknown to tourists in that region where the Alps thrust down into Provence. Nothing grew there but wild lavender.

I was crossing the area at its widest point, and after three days' walking found myself in the midst of unparalleled desolation. I camped near the vestiges of an abandoned village. I had run out of water the day before, and had to find some. These clustered houses, although in ruins, like an old wasps' nest, suggested that there must once have been a spring or well here. There was, indeed, a spring, but it was dry. The five or six houses, roofless, gnawed by wind and rain, the tiny chapel with its crumbling steeple, stood about like the houses and chapels in living villages, but all life had vanished.

A SHEPHERD

After five hours' walking I had still not found water, and there was nothing to give me any hope of finding any. All about me was the same dryness, the same coarse grasses. I thought I glimpsed in the distance a small black silhouette, upright, and took it for the trunk of a solitary tree. In any case, I started towards it. It was a shepherd. Thirty sheep lying about him on the baking earth.

He gave me a drink from his water-gourd and, a little later, took me to his cottage in a fold of the plain. He drew his water – excellent water – from a very deep natural well above which he had constructed a primitive winch.

The man spoke little. This is the way of those who live alone, but one felt that he was sure of himself, and confident in his assurance. That was unexpected in this barren country. He lived, not in a cabin, but in a real house built of stone that bore plain evidence of how his own efforts had reclaimed the ruin he had

found there on his arrival. His roof was strong and sound, the wind on the tiles made the sound of the sea upon its shores.

It was understood from the first that I should spend the night there; the nearest village was still more than a day and a half away. And besides, I was perfectly familiar with the nature of the rare village in that region. There were four or five of them scattered well apart from each other on these mountain slopes, among white oak thickets, at the extreme edge of wagon roads. They were inhabited by charcoal-burners and the living was bad. Families, crowded together in a climate that is exceedingly harsh, both in winter and in summer, found no escape from the unceasing conflict of personalities.

There was rivalry in everything, over the price of charcoal as over a pew in the church. And over all there was the wind, also ceaseless, to rasp upon the nerves.

The shepherd went to fetch a small sack and poured out a heap of acorns on the table. He began to inspect them, one by one, with great concentration, separating the good from the bad. I smoked my pipe. I did offer to help him. He told me that it was his job. And in fact, seeing the care he devoted to the task, I did not insist. That was the whole of our conversation. When he had set aside a large enough pile of good acorns he counted them out by tens, meanwhile eliminating the small ones or those which were slightly cracked, for now he examined them more closely. When he had thus selected one hundred perfect acorns he stopped and went to bed.

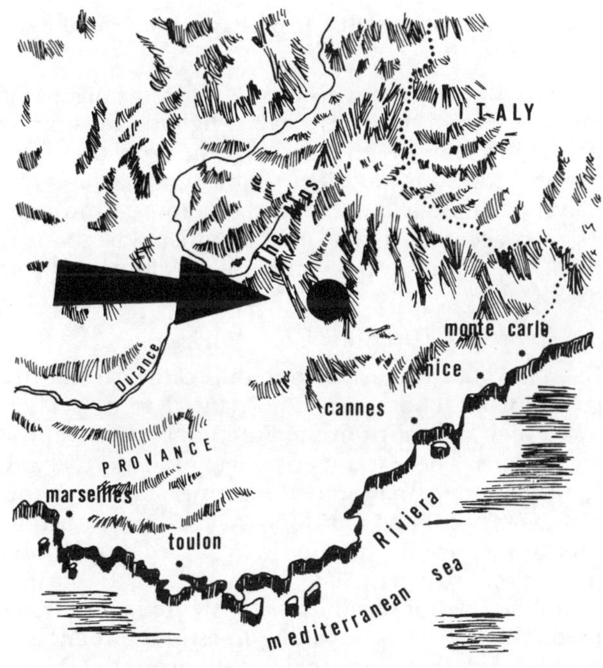

A SACK OF ACORNS

There was peace in being with this man. He found it quite natural – or, to be more exact, he gave me the impression that nothing could startle him. The rest was not absolutely necessary, but I was interested and wished to know more about him. He opened the pen and led his flock to pasture. Before leaving, he plunged his sack of carefully selected and counted acorns into a pail of water.

I noticed that he carried for a stick an iron rod as thick as my thumb and about a yard and a half long. Resting myself by walking I followed a path parallel to his. His pasture was in a valley. He left the little flock in charge of the dog, and climbed towards where I stood. I was afraid that he was about to rebuke me for my indiscretion, but it was not that at all: this was the way he was going, and he invited me to go along if I had nothing better to do. He climbed to the top of the ridge about a hundred yards away.

There he began thrusting his iron rod into the earth, making a hole in which he planted an acorn then he refilled the hole. He was planting oak trees. I asked him if the land belonged to him. He answered no. Did he know whose it was? He did not. He supposed it was community property, or perhaps belonged to people who cared nothing about it. He was not interested in finding out whose it was. He planted his hundred acorns with the greatest care. After the midday meal he resumed his planting. I suppose I must have been fairly insistent in my questioning, for he answered me. For three years he had been planting trees in this wilderness. He had planted 100,000. Of these, 20,000 had sprouted. Of the 20,000 he still expected to lose about half to rodents or to the unpredictable designs of Providence. There remained 10,000 oak trees to grow where nothing had grown before.

That was when I began to wonder about the age of this man. He was obviously over fifty. Fifty-five, he told me. His name was Elzeard Bouffier. He had once had a farm in the lowlands. There he had had his life. He had lost his only son, then his wife. He had withdrawn into this solitude, where his pleasure was to live leisurely with his lambs and his dog. It was his opinion that this land was dying for want of trees. He added that, having no very pressing business of his own, he had resolved to remedy this state of affairs.

Since I was at that time, in spite of my youth, leading a solitary life, I understood how to deal with solitary spirits. But my very youth forced me to consider the future in relation to myself and to a certain quest for happiness. I told him that in thirty years his 10,000 oaks would be magnificent. He answered quite simply that if God granted him life, in thirty years he would have planted so many more that those 10,000 would be like a drop of water in the ocean.

Besides, he was now studying the reproduction of beech trees and had a nursery of seedlings grown from beech-nuts near his cottage. The seedlings, which he protected from his sheep with a wire fence, were very beautiful. He was also considering birches for the valleys where, he told me, there was a certain amount of moisture a few yards below the surface of the soil.

The next day we parted.

AFTER THE WAR

The following year came the War of 1914, in which I was involved for the next five years. An infantryman hardly had time for reflecting upon trees. To tell the truth, the thing itself had made no impression upon me. I had considered it as a hobby, a stamp collection, and forgotten it.

The war over, I found myself possessed of a tiny demobilization bonus and a large desire to breathe fresh air for a while. It was with no other objective that I again took the road to the barren lands.

The countryside had not changed. However, beyond the deserted village I glimpsed in the distance a sort of greyish mist that covered the mountains like a carpet. Since the day before, I had begun to think again of the shepherd tree planter. "Ten thousand oaks," I reflected, "really do take up quite a bit of space." I had seen too many men die during those five years not to imagine easily that Elzeard Bouffier was dead, especially since, at twenty, one regards men of fifty as old men with nothing left to do but die. He was not dead. As a matter of fact, he was extremely spry. He had changed jobs. Now he had only four sheep, but, instead, a hundred beehives. He had got rid of the sheep because they threatened his young trees. For, he told me (and saw for myself), the war had disturbed him not at all. He had imperturbably continued to plant .The oaks of 1910 were then ten years old and taller than either of us. It was an impressive spectacle. I was literally speechless and, as he did not talk, we spent the whole day walking in silence through his forest. In three sections, it measured eleven kilometres in length and three kilometres at its greatest width. When you remember that all this had sprung from the hands and soul of this one man without technical resources, you understand that men could be as effectual as God in realms other than that of destruction.

He had pursued his plan, and beech trees as high as my shoulder, spreading out as far as the eye could reach, confirmed it. He showed me handsome clumps of birch planted five years before – that is, in 1915, when I had been fighting at Verdun. He had set them out in all the valleys where he had guessed – and rightly – that there was moisture almost at the surface of the ground.

A CHAIN REACTION

Creation seemed to come about in a sort of chain reaction. He did not worry about it; he was determinedly pursuing his task in all its simplicity; but as we went back towards the village I saw water flowing in brooks that had been dry since the memory of man. This was the most impressive result of chain reaction that I had seen. These dry streams had once, long ago, run with water. Some of the dreary villages I mentioned before had been built on the sites of ancient Roman settlements, traces of which still remained; and archeologists exploring there had found fishhooks where, in the twentieth century, cisterns were needed to assure a small supply of water.

The wind, too, scattered seeds. As the water reappeared, so there reappeared willows, rushes, meadows, gardens, flowers and a certain purpose in being alive. But the transformation took place so gradually that it became part of the pattern without causing astonishment. That is why no one meddled with Elzeard Bouffier's work. If he had been detected he would have had opposition. He was undetectable. Who in the villages or in the administration could have dreamed of such perseverance in a magnificent generosity?

To have anything like a precise idea of his exceptional character one must not forget that he worked in total solitude; so total that, towards the end of his life, he lost the habit of speech. Or perhaps it was that he saw no need for it.

In 1933 he received a visit from a forest ranger who notified him of an order against lighting fires out of doors for fear of endangering the growth of the 'natural' forest. It was the first time, the man told him naively, that he had ever heard of a forest growing of its own accord. At that time Bouffier was about to plant beeches at a spot some twelve kilometres from his cottage.

AN OFFICIAL DELEGATION

In 1935 a whole delegation came from the government to examine the 'natural forest'. There was a high official from the Forest Service, a Deputy, technicians. There was a great deal of ineffectual talk. It was decided that something must be done and, fortunately, nothing was done except the only helpful thing: the whole forest was placed under the protection of the State, and charcoal burning prohibited. For it was impossible not to be captivated by the beauty of those young trees in the fullness of health, and they cast their spell over the Deputy himself.

A friend of mine was among the forestry officers of the delegation. To him I explained the mystery. One day the following week we went together to see Elzeard Bouffier. We found him hard at work, some ten kilometres from the spot where the inspection had taken place.

This forester was not my friend for nothing. He was aware of values. He knew how to keep silent. I delivered the eggs I had brought as a present. We shared our lunch among the three of us and spent several hours in wordless contemplation of the countryside.

In the direction from which we had come the slopes were covered with trees twenty to twenty-five feet tall. I remembered how the land had looked in 1913, a desert. Peaceful, regular toil, the vigorous mountain

34

air, frugality and, above all, serenity in the spirit had endowed this old man with awe-inspiring health. He was one of God's athletes. I wondered how many more acres he was going to cover with trees.

A WAY TO BE HAPPY

Before leaving, my friend made a brief suggestion about certain species of trees that the soil here seemed particularly suited for. He did not force the point, For the very good reason, he told me later, "that Bouffier knows more about it than I do." At the end of an hour's walking – having turned it over in his mind – he added, "He knows a lot more about it than anybody. He's discovered a wonderful way to be happy."

It was thanks to this officer that not only the forest but also the happiness of the man was protected. He delegated three rangers to the task, and so terrorized them that they remained proof against all the bottles of wine the charcoal-burners could offer.

The only serious danger to the work occurred during the War of 1939. As cars were being run on gazogenes (wood-burning generators), there was never enough wood. Cutting was started among the oaks of 1910, but the area was so far from any railway that the enterprise turned out to be financially unsound. It was abandoned. The shepherd had seen nothing of it. He was thirty kilometres away, peacefully continuing his work ignoring the war of 1939 as he had ignored that of 1914.

I saw Elzeard Bouffier for the last time in June of 1945. He was then eighty-seven. I started back along the route through the wastelands; but now, in spite of the disorder in which the war had left the country, there was a bus running between the Durance Valley and the mountain. I attributed the fact that I no longer recognized the scenes of my earlier journeys to this relatively speedy transportation. It took the name of a village to convince me that I was actually in that region that had been all ruins and desolation.

The bus put me down at Vergons. In 1913 this hamlet of ten or twelve houses had three inhabitants. They had been savage creatures, hating one another, living by trapping game, little removed, physically and morally, from prehistoric conditions. All about them nettles were feeding upon the remains of abandoned houses. Their condition had been beyond hope. For them, nothing but to await death – a situation which rarely predisposes to virtue.

Everything was changed. Even the air. Instead of the harsh dry winds that used to attack me, a gentle breeze was blowing, laden with scents. A sound like water came from the mountains: it was the wind in the forest. Most amazing of all, I heard the actual sound of water falling into a pool. I saw that a fountain had been built, that it flowed freely and – what touched me most – that someone had planted a linden beside it, a linden that must have been four years old, already in full leaf, the incontestable symbol of resurrection.

Besides, Vergons bore evidence of labor at the sort of undertaking for which hope is required. Hope, then, had returned. Ruins had been cleared away, dilapidated walls torn down and five houses restored. Now there were twenty-eight inhabitants, four of them young married couples. The new houses, freshly plastered, were surrounded by gardens where vegetables and flowers grew in orderly confusion, cabbages and roses, leeks and snapdragons, celery and anemones. It was now a village where one would like to live.

From that point I went on foot. The war just finished had not allowed the full blooming of life, but Lazarus was out of the tomb. On the lower slopes of the mountain I saw little fields of barley and rye; deep in the narrow valley the meadows were turning green.

It has taken only the eight years since then for the whole countryside to glow with health and prosperity. On the site of the ruins I had seen in 1913 now stand neat farms, cleanly plastered, testifying to a happy and comfortable life. The old streams, fed by the rains and snows that the forest conserves, are flowing again. Their waters have been channeled. On each farm, in groves of maples, fountain pools overflow onto carpets of fresh mint. Little by little the villages have been rebuilt. People from the plains, where land is costly, have settled here, bringing youth, motion, the spirit of adventure. Along the roads you meet hearty men and women, boys and girls who understand laughter and have recovered a taste for picnics. Counting the former population, unrecognizable now that they live in comfort, more than 10,000 people owe their happiness to Elzeard Bouffier.

When I reflect that one man armed only with his own physical and moral resources, was able to cause this land Canaan to spring from the wasteland, I am convinced that, in spite of everything, humanity is admirable. But when I compute the unfailing greatness of spirit and tenacity of benevolence that it must have taken to achieve this result, I am taken with an immense respect for that old and unlearned peasant who was able to complete a work worthy of God.

Elzeard Bouffier died peacefully in 1947 at the hospice in Banon.

Many thousands of people have been inspired by the story of Elzeard Bouffier.

Jean Giono died in 1970, his daughter, Aline, later wrote:

"The story has had a number of titles:

The Man Who Planted Trees
The Story of Elzeard Bouffier
The Man Who Planted Hope and Gathered Happiness
The Most Extraordinary Character I've Ever Met

The story has been published in several magazines. It has always been very well recieved... Some Friends of The Trees, Nature Lovers, or Truth Lovers might be disappointed... In 1957 my father wrote a letter which explains what he wanted to achieve with his story."

Jean Giono writes, "Sorry to disappoint you, but Elzeard Bouffier is an invented character. My aim was to get people to love trees, or rather, to get them to love to plant trees... Now if I judge from the result, this imaginary character has acheived the aim... This is one of my pieces of work, of which I am most proud. It doesn't give me one cent and that's why it's achieved its purpose so well."

Jean Giono

"Everywhere in the world where the story has been published, people believed it's true. Reality meets with fiction, and one day one might indeed come across such an old shepherd."

Aline Giono

COMMENT

By the hands of Jean Giono, Elzeard Bouffier planted acorns in the soils of Vernon. By the hands of Elzeard Bouffier visions have been planted in the minds of thousands. As Elzeard Bouffier's forests flourished and rivers ran, so too will our new awareness bring hope and happiness. As history has shown, fantasy become reality, modern man's collective achievements far exceed even the wildest visions of our ancestors.

When we discovered this story to be fiction, we were saddened. We have since realised however, that hope has truly been planted. As we look around us in a more postive way we *understand that men could be as effectual as God in realms other than that of destruction*.

A copy of the letter by Jean Giono, 1957. Translated, in part, above.

PALMGROVE PLANTATION

By JACK COLES, DIANA COLES, MAX LINDEGGER and ROBERT TAP
Spring 1985

"Humanity is a part of nature, not apart from it..." The speaker is Jack Coles, an athletic man in his mid-fifties. Well developed, bronzed shoulders and muscular arms testify to years of physical work. His healthy outdoor life must agree with him – for there is barely a hint of grey to tinge his thick brown hair.

Jack and his wife Diana have just returned from a ten kilometre run in the state forest adjoining their twelve hectare property. Though they are well past fifty, there is no trace of 'middle aged spread'. Both work in an occupation renowned for its demanding toil – commercial production of bananas.

We had arrived earlier than arranged. Before our hosts' return we had strolled around in their Permaculture orchard, noting, with approval, seventy or so varieties of carefully nurtured fruit trees that must assure self-sufficiency in food, with a surplus to sell or give away. The banana plants look very healthy and well tended. Towering native trees offer protection from the worst winds. In recent supplementary plantings we were delighted to see a mixture of native trees that afford habitat and shelter for wildlife.

The house is situated out of sight, two hundred metres from the road. It nestles amongst palms, tree ferns and lovely shade trees, most of which Jack and Diana planted soon after settling here eleven years ago. We are three kilometres from Yandina, about a hundred kilometres north of Brisbane, Queensland, Australia. There is a lovely view over rolling fields set against a background of the craggy silhouette of Mount Ninderry. From the tall trees comes a riotous chorus of birds...and a flock of colourful lorikeets rockets overhead, screeching wildly as they vacate the bottle-brush tree near the packing shed.

Jack is answering our question: "What made you decide to leave a well paid and successful career in Canada to grow bananas commercially in Australia?"

Jack and Diana Coles.

Taking time to choose his words, Jack explains..."It was not an impulsive decision. It was some fifteen years in the making. Diane and I enjoy working together and being outdoors. So many activities in society divide and separate a couple. No wonder lots of marriages founder. There is so much more to life than making money that we decided to live as naturally as we could, and in a situation where we have to rely on each other. It is a wonderful way to enjoy marriage to the full."

"Incidentally, from all the reports I get, you're getting a tremendous reputation for top quality fruit, but why bananas?" Max asks. "I suppose," Diana replies, "because they were already here. We went into it pretty thoroughly, and compared to many other crops, they can be managed with relatively minimal use of dangerous pesticides and without a lot of machinery. Also, we were absolutely captivated by the property itself."

Jack continues, "We've come to feel a sort of reverential regard for this place – akin to the Aboriginal attitude. The land now owns us...it sustains us, so we have a reciprocal loyalty to that which gives us life..." He pauses...his mind seems far away, then he smiles at Diana, who takes his hand, urging him to go on. "When you feel that way," he continues, "you can't exploit...you want to cherish the land and to leave it better than you found it."

"So that is how you try to farm?" Robert prompts. "Yes," Jack says, "according to the saying 'Live each day as if it were your last; farm each day as if you will live for ever.'" Diana adds, "We like to give the plants treatment approaching what they might expect in a bananas' paradise!" We all laugh, and she continues, "They get frequent small feeds that match what gets taken away with the crop. Competition is reduced to a minimum. Pests are maintained at a low level by our trying to enlist the co-operation of every ally we can get."

Jack takes over, "You know, ninety percent of insects are actually friends. It makes our global chemical saturation blitz as ridiculous as it is costly. Added to that, despite the hundreds of life forms humanity has rendered extinct, we haven't yet wiped out a single species of insect...David Attenborough, I think."

"What are the worst pest problems?" asks Robert. "Let us introduce you to some," Jack responds, and we walk into the nearby plantation area. They are actually large herbs which propagate themselves by suckering. When a bunch of fruit matures, the parent plant dies, and the following sucker takes over in dominance. Trash from the decaying parents forms a thick mulch, spread neatly and carefully around the plants. Max remarks on the way this material is kept more than a metre away from the bananas, and that the dead leaves have been removed from them. "Why is that?" he asks. "Because of this little chap," Jack replies, rummaging through the decaying trash. "Got one, look quickly before he disappears..." He has rolled over part of a chopped up stem, and we see that it is honeycombed with tiny holes, several millimetres in diameter. A little white grub near the surface is quickly disappearing back inside. "That's the infamous borer", Diana explains, "but it's only doing what nature intended. Its part of the recycling squad." "Only when it munches its way into the growing plants is there trouble," Jack tells us, "so that's why we keep the fallen trash away. The adult beetles are exposed to our natural allies such as wolf spiders, skinks, birds, frogs and toads...and the larvae have no clear path to eat their way into what we are trying to protect. Many growers are using very powerful chemicals, such as dieldrin, twice a year to control it. By this kind of 'hygiene' we have managed to go for up to eight years

on some patches with no chemical treatment at all. Pressure from environmentalists is making people so much more aware of problems that this kind of thinking is legitimized now in *"Integrated Pest Control"*. "You didn't mention," Diana interjects, "that this is also quite effective against the nematodes that damage the roots." "Thanks, Diana. Many growers are now putting out three treatments a year to 'control' nematodes...and we have never put out any nematicides, in eleven years."

"A lot of humanity's problems are self-induced," he continues. "The devastating fungal Panama disease has virtually wiped tall varieties of bananas off the earth. Except for a few areas of this country, there are no other lady finger bananas in the world. Amazingly, the disease seems to be a plantation phenomenon. According to Stover (1972), the problem had never been encountered in the wild. Now it is spreading, even into varieties of bananas previously thought to be immune. There are currently four mutated strains of the fungus." "Are your plants affected?" asks Robert. "So far, one suspect only – but it has not spread. This plantation,

Healthy stems surrounded by trash that is kept well clear from the base.

Rows of bananas planted along the contours. Coupled with trash retention this practice minimises erosion.

wiped out. Many of their predators get axed as well, and the surviving pests have an environment with reduced competition (from their late lamented relatives) and with reduced numbers of predators! Generally, the pests have a shorter breeding cycle than predators, so they build up resistance faster! Thus, all over the planet, we are producing super-rats, golden staph, new strains of malaria or V.D. – as well as resistant moths, fleas, ticks, grubs, worms, etc. Whenever we interfere drastically, there are all sorts of violent and unforeseeable chain reaction responses such as outlined by Rachel Carson, when untargeted forms such as birds, fish, animals, even humanity are stricken."

Robert shakes his head. "Depressing isn't it. Going back to something you said earlier. You don't sound very happy about monocultures, Jack. What are the alternatives?" "Max could answer this better...and I'm sure your book will do much to reply to the question. In short though, perhaps a larger scale permaculture, with more varieties of foods, in smaller concentrations, with companion plantings, readiness to accept some losses from pests. Why must humanity be so greedy as to demand more than a hundred percent from the land? We should not be taking out more than we put back. A sustained yield will last forever, but too many human activities destroy the very source – like in Aesop's fable of the goose that laid the golden eggs. Also everybody should be encouraged to grow more of their own food on their own properties."

"Diana and Jack have already halved the area under bananas and planted thousands of trees;" Max tells Robert. "How about we go look at some of them before we leave?" As we walk, he continues, "We've already done a couple of video sections on their re-afforestation, and they've hosted a number of field days." "How many trees?" Robert asks. "Around three thousand," Diana answers, "which would be about the number planted by any of the local shire councils annually. Apart from a few weeks of casual help from outside, we have done all the planting and maintenance ourselves." Most are native trees, and many are either rare or endangered, such as kauris and red cedar. There are also hundreds of food trees for wildlife – such as lilly pilly, guavas, edible palms, mangos, banksias.

Previous owners had already established the basic pattern of tree corridors and shelter belts. Everywhere we can see evidence of our hosts' efforts to improve and enrich the variety of trees, as well as give an air of tropical enchantment. Even tree stumps have been planted with orchids. Small groves of palms are found everywhere.

by the way, is the longest established in this part of Queensland – it is more than thirty years old. Yet the working life of many commercial operations is as little as five years. There must be some comparative factors worth researching – surely it cannot be luck entirely?"

"Very interesting," Max comments. "Getting back to your views on pesticides, I know you have concerns other than residues in food, or runoffs...would you like to talk about resistance?"

"For those who may be unfamiliar with resistance, it is believed to develop roughly like this: the shared gene process of reproduction is a very successful biological process ensuring diversification of the characteristics of living things. This enables more biological niches to be utilized. Thus, when you blast out a currently 'effective' pesticide, there is a variety of responses from the targeted creature. Most individuals on the receiving end are wiped out. Since the major part of the breeding stock is no longer in evidence, the user is well satisfied – BUT those pests that survived, did so because they had a genetic predisposition to be less affected by the chemical. It is almost as though nature had foreseen certain possible contingencies that could threaten the species and had already taken steps to ensure survival of that life form even at the expense of its having to change some of its characteristics. These survivors re-group and breed. They all have a higher probability of passing on their resistance to their offspring."

"Of course, it is not only the 'target' species that gets

Some of the trees planted by Jack and Diana. These Eucalypts now form part of an essential windbreak and contribute to the wildlife corridors.

Palms and tropical enchantment.

The new trees, some of which are now eight years old, are, in places, up to twenty metres in height. In a once badly eroded gully, now reclaimed, we notice that nearly every tree has a birds' nest. "Why such a concentration here?" Robert asks. "We are pretty sure that it is because these are spikey sorts of trees, such as bunyas, hoops and kauris. The spines help to stabilize the nests, and we think that they keep predators at a safe distance," Diana answers.

"Why did you decide not to plant exotic pines, most people would, don't you agree Max?" "Not only do pines ultimately sterilize the soil, according to Richard St. Barbe Baker – but they are completely unsuitable for native wildlife," Jack answers.

As we head back to the house, Diana tells us she will go home by a shorter route, as she wants to put out the 'wallaby bananas.' She explains how they put out rejected fruit in the forest for the wildlife.

We toil back up the hill to the house, listening to Jack talking earnestly about the importance of neighbouring properties establishing tree corridors for wildlife – and suggesting that it could be co-ordinated by an Environment Ministry.

COMMENT

Our visit with Jack and Diana was one of inspiration. So many of us strive to succeed, for financial rewards. Here a couple have achieved such rewards almost as a by-product, of their reverance for themselves as individuals, as a couple in marriage, and as a working team, and for the earth environment in which they live.

REFERENCES

Carson, Rachel
Silent Spring
Houghton Mifflin Co. U.S.A.

Stover, R.H. (1972)
"Banana, Plantation and Abaca diseases".
Commonwealth Mycological Institute. England.

THE HONNEF FAMILY FARM

By FRANZ HONNEF

The Honnef family farm represents 5 hectares of Permaculture productivity. The farm is located 20 km north of Brisbane (27.10 S, 153.00 E). Within its borders can be found the following elements of a finely tuned system .

5 bedroom family home, 7 poultry sheds (26,000 broilers) fenced into 6 paddocks, 3 bulk feed bins, 2 dams (with fish), well and wind mill and town water. Extensive irrigation supplying water to 54 different varieties of tropical fruit and nut trees including 1,142 macadamia trees, 700 banana trees, 200 paw paw trees, over 100 various varieties guava trees, 70 custard apples, 28 different citrus fruits, 40 Brazillian cherry trees, 12 mango trees, 235 tamarillo, 20 avocado trees, 10 coffee plants, over 50 different palms, 25 tamarind trees, 40 grape vines, 5 Indian walnuts, 5 jackfruit, 9 Chinese raisins, 9 mulberries, 7 bean trees, numerous pea trees, 4 apricots, 2 peaches, 2 Japanese plum trees and olives. Others include roseapples, river cherries, native tamarind, longans, tung nuts, horse radish tree, kiwi vines, various passionfruits, five corner fruits and quite a comprehensive list of numerous other exotic fruit and nut trees.

When I came here in 1974 it was the Brisbane flood year. It rained from November to June, everything was wet and soft. Most of the land was planted with very poor native shrubs and some big sugar gums. First I made a plan. It took me a long time to work everything out. I had no contact with Permaculture. I'd never heard about it.

My main teachers were the natives of East New Guinea. I had spent eleven years with them living in their villages and making big gardens with them.

So first I started fencing, to keep every unwanted living thing outside and every animal which belonged to us inside. I divided the land into small blocks for better control. Then I started to work on one block at a time. We cleared the lantana and scrub and I started our nursery for nut and fruit trees. As the land was cleared we planted macadamia nuts and big mixtures of nut and fruit trees to have nuts and fruits all the year round. I discovered for myself that the best nitrogen fixer and fastest grower is the *Cassia siamea*. It does very well in our local area. Macadamia nut,

tamarillo and bananas grow so densely on some blocks that it looks like a rainforest.

Then I built a well on our land, bricked it to a depth of thirty-six feet and put a windmill over it to pump water for our house and garden.

For food plants I chose, like the natives, perennials. First comes sweet potatoes. We boil the tips for spinach and the potatoes are always in the ground ready to be picked up. So are the paw paw pumpkins which are true perennials on our farm. Chokos are vegies with a hundred recipes. We eat the tips for greens and the tubers about three feet down are excellent to eat. Next come the gourds. Wax gourds grow to 40 kg and store for five months. Many people in Narangba have wax gourd for dinner! New Guinea snake bean grows two metres long and fruits heavily. Guada gourd is a perennial. Indian pea tree is a very good food. The lab-lab niger (hyacinth bean) is a wonderful green bean and the New Guinea sword bean is grown everywhere here to supply us with dried beans all the year round. The luffa is a delicate gourd to eat young and the sponge of the dry gourd comes in handy on the farm. All this food grows without any maintenance, outgrows weeds, and needs no pest control.

I grow turkeys for Christmas on the farm. They are semi-wild, sleeping in the wattle trees and cleaning and fertilising our land. Other animals include king pigeons, bantam chickens, game fowls, geese, ducks, goats (for milk) and bees for honey. Our energy is solar for hot water (self-made), wind for pumping water, and firewood (we never run out).

I grow Indian black maize, barley and oats, all the Fukuoka way. It works well. For fresh green vegies we eat many weeds. Every season has its own wild plants.

For nitrogen fixing plants I use *Cassia siamea, Cassia alata,* Brisbane wattles, bean trees, Indian pea tree, leucaenas, *Sesbania formosa,* acacias, some ornamentals, jacaranda and many more all mixed between nut and

fruit trees to create a forest system containing useful animals.

Our house stands in the centre and a fifteen feet wide road runs from one end of our land to the other, the full 630 metre length.

Three years ago I came in contact with Permaculture Nambour and it was our big blessing. Our life here on our farm is joyful. My wife Erika and our seven children are all in it as one team. We find great joy and satisfaction in doing it our way, living a healthy happy life with God and his creation in harmony and peace.

CITY FARMS

By KAY KNIGHTS

Cattle rustling in Newham, sheep dipping in Inner London, moving pigs into the centre of Newcastle; City Farms do have similar problems to real farms even if they are different in concept. Perched above the railway tracks in North London lies a tiny farm where barnyard animals are cared for by city kids. City Farm 1, the forerunner of the new City Farm Movement, was started in 1972 on the site of an old timber yard and a British Rail shunting depot. Out of the wasteland, the locals created a farm atmosphere rarely felt in an inner city neighbourhood. The enterprise was initiated by Inter-Action, the well known community resource organisation.

SUCCESS OF CITY FARM 1

City Farm 1 was an enlightening success. The project demonstrated how derelict sites with no immediate commercial value can be used to put life back into a community. The project was sparked off quite spontaneously when a group of families decided that they had at last found some space for gardening. With the co-operation of the landowners, British Rail, local people cleared the land and began growing vegetables. Eventually, they formed an allotments association which became a self-managing group. Meanwhile, the pensioners were getting in on the act on an adjacent piece of land and set up their own gardening club.

A barnyard and an indoor riding school were later developments managed by a team from Inter-Action including qualified riding instructors. Sheep, pigs, chickens and goats were reared by local children. Daisy, the Dexter cow, and Gilly, an enormous, friendly Anglo-Nubian goat, have become permanent members of the family. The project has run for five years (1972-1977) and was financed through earnings, donations and gifts in kind. In an area well known for its troublesome teenagers, vandalism was minimal. This was attributed to the intensive use of the site by people who lived close by and the fact that the site had been transformed by volunteers using seconhand materials which, having a rough-hewn finish, proved unattractive targets for vandalism. Perhaps more significant, though, was the inclusion of children and young adults in the work programme. Children had something live and real to care about. With the stables and 11 horses to care for as well as the barnyard animals, there were never enough paid staff to cope alone. The reality of dependence on the young for help meant they shared in the responsibility for the care of the animals and came to view the farm as theirs.

THE ROLE OF INTER-ACTION'S ADVISORY SERVICE

Through its Advisory Service, Inter-Action is now involved with alternative approaches to adult training, youth education, unemployment and sport; everyday problems, which in the city are compounded by physical decay and social imbalances. After many years working in the field, Inter-Action believes that ordinary people can improve their environments through their own creativeness applied to practical projects. With the proper tool and a sympathetic framework, they have found that passivity can be transformed into action. In much of its work the Advisory Service plays a catalytic role; it aims to provide the know-how and encouragement, and this is particulary evident in its approach to the City Farm Movement.

City Farm projects are low cost resources providing training, leisure and educational activities for the elderly and the handicapped as well as the able young. Following on the success of City Farm 1 and a visit to the site by Department of the Environment officials complete with "wellies", Inter-Action received a Government grant. Available for three years, the grant enables three members of Inter-Action to assist other voluntary organisations to set up similar projects on the derelict sites that pepper our inner cities.

City Farms are unlike real farms because they are not part of an industry concerned with the commercial production of food. Their role could be considered introductory to the concept of real farming, allowing city people to experience the pleasures and hazards associated with being "down on the farm", a place where the house cow or the backyard pig can become the responsibility of a small group of adults or children for the benefit of their neighbourhood rather than just an individual family. City Farms do not become self-sufficient through produce, but they can supplement their budgets through a social and educational programme introducing city people to animal husbandry and horticulture.

The practical nature of City Farms depends on the character of a neighbourhood and the restraints which may be imposed by wasteland sites. Mostly, children and adults want to experience activities that they have only seen or heard about: simple things that are basic to existence outside the normal city environment.

Inter-Action encourages the local management of City Farm projects in the belief that the current physical decay and social need prevalent in cities are best tackled by those who have their roots, their families and their futures in the neighbourhood. Responsibilities for management are matched by the satisfaction of planning to suit local needs and reaping the benefits in kind. Kids can run a chicken club and collect the eggs; older boys can build rabbit hutches and supervise the sale of small animals to local people, while the elderly can grow their vegetables and keep the produce. In neighbourhoods where the old community has been broken up in order to make way for the planners' demands, new projects which involve the whole family become necessary to reinforce what social life remains. People are motivated by city farms because gardening and the caring of animals is as natural to city folk as it is to their country cousins; they simply need the land. If projects are to be sustained, local management providing tasks for all sections of the community is the key.

Whatever the scale of a project each has to go through similar hoops. Bureaucracy is not very kind to novices, but it does set up a useful discipline. The Inter-Action Advisory Service shows groups the reality of the problems they will face and helps them think them through. The group must be able to think about the running of the farm once they are past the setting up stage. Respect for each group's individuality is important if relationships are not to be undermined and some groups approaching the Advisory Service are already skilled.

ESTABLISHMENT OF CITY FARMS IN BRISTOL, NEWHAM AND NEWCASTLE

Windmill Hill City Farm in Bristol, Newham City Farm in East London and Byker City Farm in Newcastle were among the first to approach the Advisory Service in the summer of 1976. Windmill Hill was already a skilled environmental pressure group while Newham had the services of a local clergyman who was able to carry out much administrative work in the early days and make formal applications for government and private sector grants. Many City Farm Groups have been helped by clergymen who are good intermediaries with administrative and social work skills. Byker suffered many initial management problems caused

by committee members who were unrealistic about their commitment to the project and outside its immediate catchment area. With the help of the Advisory Service, a strong local management committee has now been elected and its members are gradually proving themselves to be very capable at running affairs in which they have a stake.

Newham and Byker both sponsored Job Creation programmes, Government backed programmes employing youngsters who are out of work alongside adult supervisors. The initial poor management at Byker resulted in several workers leaving the project while, in contrast, Newham ran an excellent programme with young people building animal houses, fencing and roads with great success. Both projects have proved the usefulness of clearing away the initial on-site problems, easing a path for local volunteers to take over. Sponsorship of this kind has to be carefully weighed up as it may be too onerous a task for self-help groups who will have a moral responsibility for training and follow-up when the young adult leaves. Windmill Hill in Bristol also considered Job Creation, but after 10 months of hard negotiating with their local council they began the work using volunteers, led by an unemployed teacher.

SETTING UP A CITY FARM

A feasibility study and public consultation are the first steps in the setting-up process. The City Farm branch of the Advisory Service is able to finance feasibility studies where these are necessary, and is often called

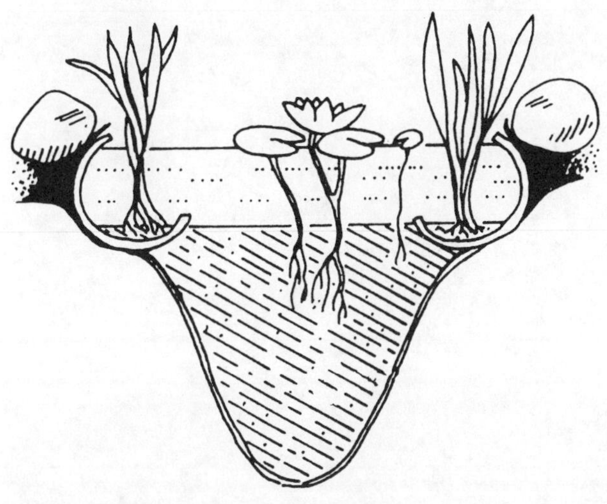

Opposite: Bill Mollison during the construction of a tyre pond at Windmill Hill City Farm, Bristol.

Above: The tyre pond – water plants have their roots in soil. A polythene liner keeps the unit water tight.
Photo and diagram, City Farm News (NFCF).

upon too, to assist with public consultations procedure and the follow-up to put ideas into action. Various techniques are employed; slide shows, discussion meetings, films, socials and special events. A slide show has been compiled by Annette Drago, a member of the City Farm team, which is able to show the achievements of different groups around the country and is now much in demand. The Department of Education and Science is taking a copy to Russia.

From the Inter-Action Community Resource Centre in Kentish Town, groups are able to use slides, films, printing facilities and video equipment for film making, sometimes in conjunction with the media van. This large, red, converted Mercedes-Benz van – a cinema and TV studio on wheels – never fails to arouse interest in even the most apathetic neighbourhood. Driving into housing estates or shopping centres, by playing music, showing films and encouraging discussions between local people on the spot, it has a Pied Piper effect which enables the group to reach and consult hundreds in an afternoon. Obviously, social events tend to be useful forerunners to the more traditional methods of consultation like public meetings, and are sometimes more effective in obtaining commitment and ideas for a project. For instance, imagine a derelict site transformed overnight to appear like a "farm". A living exhibition with signs and mini-stage sets depicting rural scenes like duck ponds, nursery gardens, animal paddocks and residents in appropriate dress. An air of sober authenticity is provided by a local farmer and his animals and the whole affair is supervised by the mounted police. Windmill Hill in Bristol did this and 1500 people gave them their support, which has been sustained ever since.

Groups are encouraged to draw on the services of their own local specialists and to try to set up a group of advisers who would be willing to donate the odd hour or two to give advice. Farmers, businessmen, vets, solicitors and builders have helped various groups. They derive great satisfaction from applying their services directly to a community group and, usually being residents, they, too, can benefit. The importance of links between established professionals and self-help cannot be overestimated. The National Farmers' Union has given its blessing to the City Farm Movement. The community at large will surely benefit from a better understanding between town and country.

Clearing derelict sites, fencing and building can be an arduous task, but self-help groups do have their advantages. The cost of labour and materials is usually much lower than a commercial budget would allow as demonstrated at Windmill Hill. Cash donations or donations in kind are often found and many groups have been able to tap the secondhand and waste materials markets. A good "hustler" is essential to every self-help group. Local agencies and official bodies can be helpful; the Probation Service, the Scouts, the Army, Council Amenity Departments, Community Industry, the AA, large contractors and others from the private sector have all shown a willingness to donate either time or materials. Gifts of unwanted sand, top soil, end-of-the-day concrete, timber and perhaps more important the transport to collect them have been obtained by the more enterprising farms from areas with rich resources. Not all are so lucky, but the re-use of waste from industry on an organised basis will become increasingly important in the fostering of self-help groups if they are to realise their full potential.

RELATIONS WITH THE AGRICULTURAL INDUSTRY

Over the next few years as more City Farms embark on educational projects, relations with the agricultural industry will become even stronger. The National Farmers' Union of England and Wales, the Association of Agriculture and the Department of Agriculture of the Aylesbury College of Further Education have been the first to join with British Rail and the Department of the Environment in lending their support to City Farms through the Inter-Action Advisory Service. This kind of backing is essential to the development and training enterprises which have no other support group.

Self-help groups are a valuable asset to any community. They can provide for essential social and educational needs at low cost and without too much extra burden on ratepayers' money. In addition, their role has an inbuilt mechanism for strengthening links between all sections of the community and thereby fostering understanding of mutual problems. As urbanisation spreads, so the understanding between town and country becomes more remote. In the city there is no distance between tower block, terrace and the patch of wasteland and people are already taking the opportunity to plant the seeds of a new understanding themselves in their own backyards. The City Farm Movement is developing and projects already exist in Birmingham, Bristol, Erskine, Glasgow, Liverpool, London, Milton Keynes, Newcastle, Nottingham, Sunderland and Swindon.

National Federation for City Farms (N.F.C.F.) 66 Fraser Street, Bedminster Bristol BS3 4LY.

WILD SCENES BY THE RAILWAY

By MIKE ROTH

We all know there's not enough green in the city, but as a matter of fact Brixton isn't 100% brick and tarmac. There must be something like 2000 trees within a half-mile of our house, counting the small park behind us, the trees in the gardens and those lining the triangle of once elegant streets of which ours is one side. And quite a mixture too, including a couple of old mulberries, some white willow and lots of robinia. There are even fruiting apple trees on the street.

But the one and only stretch of wild woodland is a four acre strip running along one stretch of the spider-web of railway lines here. None of the trees looks more than thirty years old – but they do look as if they've planted themselves. There's a pleasantly unornamental feel about the higgledy-piggledy clumps of birch, oak, sycamore, lime and elder which are the majority of the tree species. Likewise there's the sort of wild life you might expect in an undisturbed, self-seeded area of woodland, including foxes and a variety of birds and insects.

The land is at present the subject of a stalemate between British Rail, the owners, who see it as a light industrial site and a saleable asset, and local residents and a council who would certainly block any plan which would reduce a substantial area to concrete and noise. We are hoping the council will resolve the uncertainty by buying the site and licensing or leasing it to us. Our plan, for which the council has agreed to sponsor a feasibility study, is to preserve as much of the wild plant and animal life as possible, while developing the least wooded end of the site as a garden centre, to specialise in wild and traditional native varieties of herbs, flowers, vegetables, shrubs and trees. At the same time we would be developing the woodland edge to provide yields of fruit, vegetables and nursery plants, and also making pathways through the denser part of the wood. The overall intention would be to provide an amenity of beauty, educational interest and practical use which would be open to the general public. To a great extent it would still be a self-created ecology, a very rare beast indeed in a built-up area.

We are working on a design for the site at the moment, and plan to include one building: an information centre/site office combined with a solar greenhouse. We would very much welcome any advice and suggestions on any aspect from readers – and would also be happy to put anyone up for a day or two who wanted to see the site and talk over ideas for it. You can phone us on 01-274-3464 or write to 6 Loughborough Park, London SW9 8TR.

COMMENT

Mike, Judy and Ella write,
"After an initial spurt following the course at Blencarn, the spreading of Permaculture over here is continual, but fairly slowly it seems. I guess we're hampered by having no real experts over here, or any well established examples of Permaculture in the land, – as yet anyway."

TYN Y FRON – DESIGN

By PENNY STRANGE

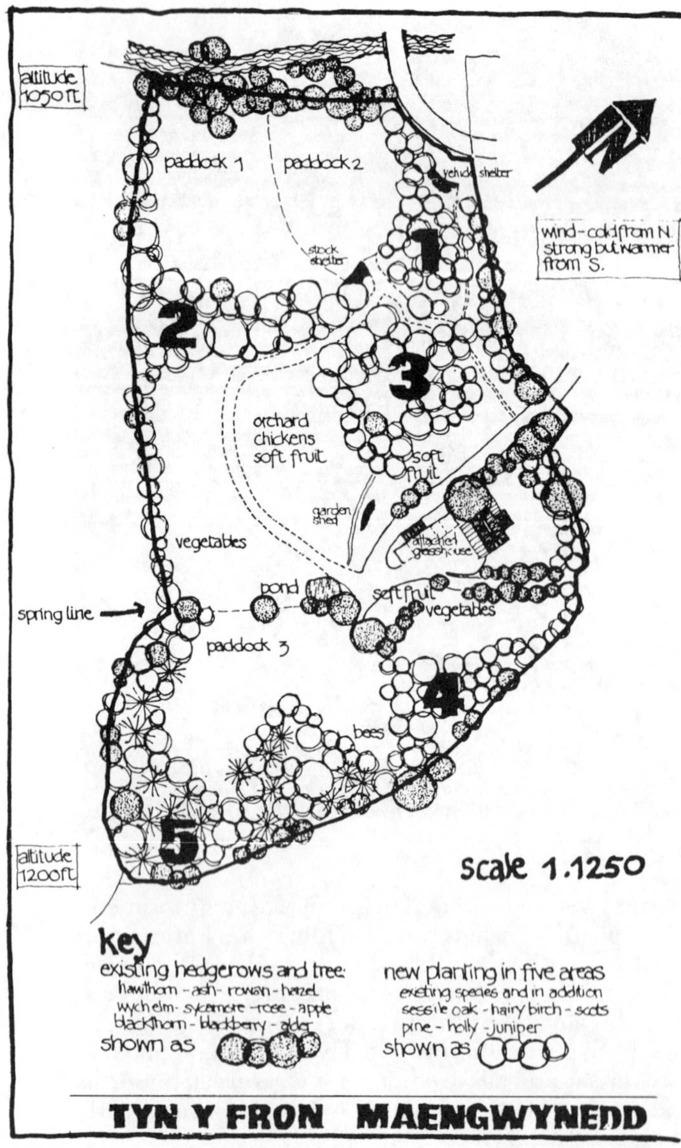

Tyn y Fron is a six-acre holding on the west-facing slope of a valley in North Wales. All the land around us is grazing for sheep, sometimes cattle. In our valley there are a fair number of trees, in hedgerows, in areas where sheep are kept out (e.g. by gorse) and in some derelict woods where the valley sides are very steep. We three adults and two children moved here in July 1983. We have been slowly working out our design, and were helped by people who came for the design weekend in September. The following is our design in progress, which we expect to modify as we go.

These are our main considerations:

SLOPE.   Some of the land is very steep, and nearly all of it is sloping.

WIND.   Strong and cold from the North, strong from the South too.

WATER.   We get lots of rain but need a water supply to areas where we may be planting out in dry spells. There are also some wet areas, and several springs, but only one is reliable all year round.

CONSERVATION.   Parts of the land are rich in flora characteristic of old grassland.

OUR OWN REQUIREMENTS.   We want to grow most of our own food and provide for our energy needs (there are no main services). We also need an adequate cash income for other expenses.

The general idea of the design is to plant trees which will divide the site up into different areas. These trees are placed to give shelter and will also provide fruit, perhaps nuts, forage for chickens, or fodder such as leaf hay, fuel and eventually timber. Each of the five numbered areas has a different mix of species to suit the situation, based on native species and in particular those we find growing in the locality. (E.g. Scots Pine, Rowas, Oak and Juniper in the high exposed area 5; Alder, Ash, Hazel and Willow in the wet area 3.)

In November 1984 we planted 4,000 young whips. We got heaps of back copies of the local community newspaper and, tearing it half across, put one round each tree. This was invaluable in marking where they are, will reduce the weed problem, and make it easier

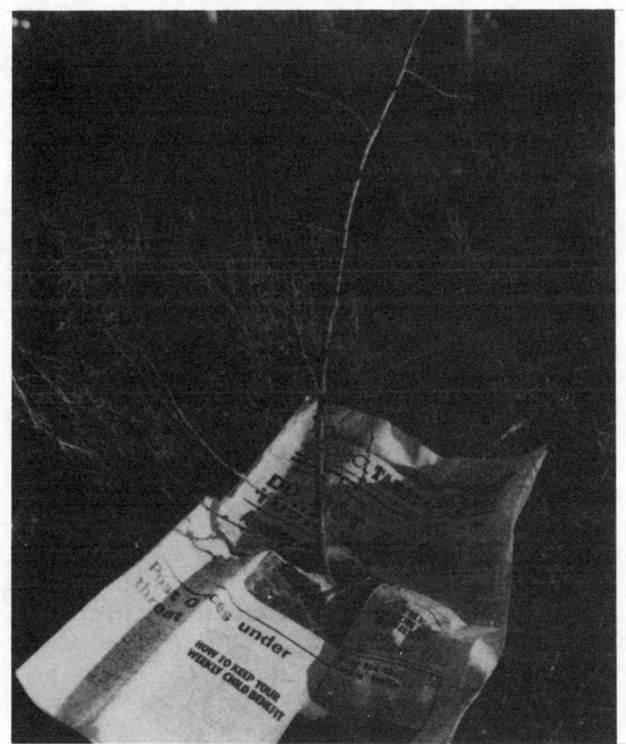

Newspaper used to suppress weed growth. (Photo, Cordelia Weedon)

boundry hedge is alder and birch. We used bracken mulch last year. The brassica patch was a weedy field: beds were marked out, and covered with manure and a thick layer of bracken during winter, and planted with cabbages in June and July. Despite the drought, they did well, and weeds and pests were minimal. It was noticeable how cabbages in unmulched parts of the garden were attacked much more by caterpillars.

CHICKENS – will be used to clear the ground and manure a different part of the garden each year. They will also free-range outside the garden. Bushes and greens for chicken forage will be included in the tree planting and orchard areas. We are considering integrating the chickens' winter housing with the glasshouse. Bee forage is another factor in our choice of tree, bush and flower plantings.

THE PADDOCKS –   are for a small flock of sheep. They include the most interesting areas of grassland which we want to keep from scrub encroachment.

to identify the trees for weeding. Some have blown away in high winds, but most have stayed and are now soggy.

THE HOUSE AND YARD AREA –   will include space for play and clothes drying etc. There will be an attached glasshouse to the south of the house, and a compost toilet somewhere near – we are looking for a suitable slope which will allow the toilet to be used on one level and emptied on a lower level, so as to avoid lifting work. We shall rebuild the outbuildings for workshop space. There is a steep bank outside the house suitable for a herb garden.

THE VEGETABLE PLOT –   occupies the only areas that are level or gently sloping. It will be sheltered by the trees on the rest of the site, and there will be hedges immediately round it and dividing it up. The hedges will be productive, providing fruit such as raspberry and blackberry, chicken forage, or mulch. We have made a start by planting alder, which provides shelter and mulch, and has nitrogen-fixing ability. As well as the usual range of vegetables we want to try growing grain in small plots for ourselves or for chicken fodder.

So far we have planted a thick hedge on the orchard side of the garden, beech, hazel and ash on its north face, raspberry and blackberry on the south. The

Penny and the children on their first visit to Tyn Y Fron.

49

The sheep will be moved from paddock to paddock so that the grass can be rested and the pest incidence reduced.

The paddocks – not for sheep but for a pony and a house cow, the pony to work, mostly carrying things, and the cow for dairy produce.

There is an existing pond and we have made two more in the wettest area, and have planted round them with willow and alder. Ponds are for ducks, and a source of mulch. We have to investigate and try out further plant and fish species. From the existing pond it is possible to make channels to provide water for the main vegetable plot. Mulching should reduce much of the need for watering in any case. There is a lot of bracken which could be cut for mulch, but as cutting and gathering it is very time-consuming, we shall extend the mulched area only slowly.

The one reliable spring is about 35 feet lower than the attic of the house, and 100 feet behind it. The water is to be pumped up to the house by hydraulic ram. At present all water has to be carried.

Our heating and a lot of our cooking relies on wood. There is plentiful dead wood locally, but we plan to supply some from our own trees in the future by coppicing.

We have not yet looked into marketing for our potential cash income, but we have been told that there is a good market for organically produced vegetables, and a local group willing to co-ordinate growers. Other sources of income might be: spinning, jam (soft fruit grows well here), coppice crafts, some workshop product or repair. We could sell free-range eggs, or keep goats and make yogurt and cheeses, but we want to sort out the basics first, before we consider the demanding business of dairying. None of these various things would make a fortune, but we don't need much money.

We sold our surplus vegetables easily last year, and next season will be selling more. We have been asked to supply bedding plants to a local shop. One of us is earning money signwriting.

THE RETENTION OF ENERGY IN A PERMACULTURE SYSTEM AND THE USE OF EDGE AND PATTERNING IN DESIGN

By KIM CHRISTIE

A Permaculture system endeavours to rely as little as possible on outside energy sources and to maximise the use of renewable resources available within the system. The aim of design should be therefore to set up the most effective possible web of living systems and structures, to catch and store as much energy as possible between the source and the point of inevitable loss from the system.

Sources of energy and resources available to a system may be sun, wind, rain, earth and even time and space, all of which should be considered in the creation of a design.

Energy storages provide yields

Plants –   Food, fibre, timber, mulch, climate modification

Animals –   Food, fibre, manure, can be used for harvesting other products of system

Water –   Stored for re-use, to redirect light, provide a habitat for food producing animals and plants, climate modification

Earth Resources –   Clay, rock for building, soil. The basic resources together with light and water for all living systems, earth resources used for banks, swales etc. For water storage and flow control, sound and wind deflection.

As can be seen, a system can provide an extremely complex web of interdependence to maximise energy within the system before the final loss. Storage should not only provide their own needs interdependently but also the needs of the people.

DESIGNING AND REGULATING THE SYSTEM

Observe each item within the system and see how many purposes it can fulfil.

For example, Groundsel - instead of eradicating it, try the following uses:

1. Existing stand can be partially cleared to provide sheltered pockets for tree establishment.

Waterflow control – keyline irrigation channel.

2. Plants can be cut at about 2 metres and fed to goats which eat leaves and twigs.

3. Remaining stems can be used for brush fences to provide shelter, contain poultry etc. or used for trellises for climbing annuals, also dried and used as small firewood.

4. Cut and shred for mulch and compost if cut before flowering there are no reseeding problems and butts will coppice providing a renewable resource - all this from a noxious weed.

Chicory and comfrey can be planted in bands across slopes in areas receiving intensive feeding to trap and use nutrients drifting down-slope. The plants are cut and fed to livestock or composted and the nutrients started back at the top of the system. Light foliage leguminous trees can be spaced through vegetable gardens/orchards (eg. leucaena, wattles) not only to fix nitrogen but to trap leached nutrients which are reused in the system by lopping and shredding the trees for mulch or compost goat or cattle fodder (30% maximum) or natural leaf-drop.

All systems should have as long a life, and require as little maintenance as possible.

51

All natural systems develop in spirals although this may not be obvious on the scale of our observation, eg. a wind appears to be blowing in a straight line, but is actually air spiralling into or out of a high or low pressure system. Physical barriers may prevent the full spiral forming but the tendency is always there.

A spiral has its own dynamic ability to concentrate forces and thus increase yields, a curve being a part of a spiral also has some of this aspect. Crops planted in curved lines have been shown to provide up to a third more yield than ones planted in straight lines.

Another natural phenomenon which can be used to increase yield is 'edge'.

EDGE on a large scale is where two dissimilar ecosystems meet, but may also be as small as the line along a fence or a raised garden. It is a zone which differs from the two adjoining zones.

EVENTS at edges can differ dramatically from adjoining zones -flowering and fruiting patterns can be brought forward on a sun facing edge or retarded on a shady side.

Edge combined with curves or spirals can multiply available microclimates, thus events and appropriate species and therefore yield.

The old-style farm with small hedge (edge) rowed fields was far more productive per unit area than modern large open farms designed to allow for mechanisation.

ACCUMULATION Yields and energy which are in transit through a system tend to accumulate at edges, eg. seeds from plants, nutrient wash, plant debris, birds and animals tend to live in the dominant systems but feed and nest in the edge zone.

PRACTICAL USES The yield of grains can be as much as doubled for two metres from an edge, this edge can simply be an adjoining beneficial companion crop. This peculiarity can be utilised as follows:

The yield can be further increased by curving the plantings across the field.

Remember also in laying out curves to try and use them for other purposes than just yield increase and creation of microclimates.
- To channel or deflect wind
- To channel or deflect cold air traps in sub-tropical areas for tropical peaches and nectarines, kiwi fruit etc.
- combine with contour lines to deflect water flows, reduce erosion

SMALL AREAS Small garden edge can be created by curved and tiered gardens, sleepers or concrete blocks.

A mere 2 metre diameter raised spiral provides over 16 metres of linear planting with varied microclimates.

A final point to remember is that there is no limit to yield except your imagination - something more can always be added.

Coral tree cuttings will soon take root and become a living fence; so defining an edge.

Choko vine, leucaena and paw paw form an edge and circular windbreak – creating a microclimate within.

THE PLACE OF WEEDS

By DAVID HOLMGREN

I would like to make a few observations and suggestions concerning certain weeds – bracken, wattle and blackberry. These plants are common throughout Tasmania and parts of mainland Australia, invading badly managed or abandoned farmland, logged and fired bush. What I am trying to get to is an appreciation of the ecology of weeds. From this we can learn not to hate a plant just because it gets in our way and possibly how to use it to advantage.

Weeds have been described as plants out of place (where we don't want them) which says something about us but nothing about the plants. I would say weeds are pioneer species which colonise disturbed habitats. Since disturbance (flood, fire, landslip, volcanism, etc.) is a part of nature, certain plants have evolved to recolonise affected habitats and it is these species which comprise most of our weeds. "Non-weedy" species can become weeds when introduced to a new environment because natural limiting factors such as parasites are not present. Man has probably been the largest single cause of instability mainly through forest clearing and burning. It's appropriate therefore that a whole array of species has evolved specifically to cope with human created habitats.

So it should be possible to recognise weedy characteristics in plants independent of whether or not they are causing us any problems. Weeds tend to have the following characteristics. These are relative factors and are in no way definitive.

(1) Short lived relative to species of the same plant type, e.g. silver wattle compared to blackwoods or Eucalypts: bracken compared to king fern.

(2) Abundant reproductive capacities by seeding or vegetative reproduction, e.g. thistles. This allows unstable areas to be quickly colonised.

(3) Nectar and pollen sources for bees. Most weeds provide some flow and many are renowned, e.g. patersons curse, black locust (weed tree of N. America). This is an adaption which encourages a high seed set.

(4) Fast and vigorous growth. This helps a quick recolonisation.

(5) Capacity to handle very poor, compacted or leached soils.

Many weeds are nitrogen fixers, e.g. gorse and broom. Others are specifically adapted to humus free soils, e.g. Mullien. Other characteristics could be noted.

So we can say that in Tasmania, bracken, blackberry and silver wattle qualify very well as weeds in our ecological framework. What then are these weeds up to all over our countryside?

Firstly, the much despised bracken: It's quickly spreading rhizomous root system stabilises ash (fired) soil, soaking up the soluble nutrients before they are leached, preventing erosion and building up the humus with its copious fronds which die off each season. It is assisted in its spread by being unpalatable to almost everything. Large pure bracken stands provide little food for browsing animals. This means the animals go elsewhere and tree seedlings get established (not very light demanding species such as Eucalpyts) in a moist, sheltered, frost free environment. Being fairly light demanding, bracken dwindles to a few scattered fronds in a well developed forest. Sounds ideal. Of course bracken is a great fuel accumulator, so although its very growth leads to its elimination, it plays at encouraging fire which will regenerate it. Can you blame it?

How about the silver wattle? It comes up after fires too, but also germinates under established bracken or even in pasture. In the native ecology it could be placed between bracken and Eucalypts (the Eucalypts always regenerating at the same time as the Acacias but eventually succeeding them) but this tree is very versatile in form, habitat and relationships. It builds up the humus very quickly and is a nitrogen fixer. Fast grower? I've seen it 60 ft. high at 10 years old on deep moist clay loams in the Huon, which rivals any of the world's "weed" trees such as pines and poplars. The silver wattle flowers profusely, a source of pollen for bees and sets seed accordingly, which can lie dormant in the soil for years until required to burst into life. And the blackberry? – that introduced noxious weed of which in 1895, Baron von Mueller said "deserves to be naturalised on the rivulets of any range". This spiney bramble controls erosion especially along streams which have been de-stabilised by land clearing in the headwaters. It has deep roots and is partly deciduous, acting as a nutrient pump, bringing minerals back to the surface, depositing them as humus: It is excellent bee forage and supports large blackbird populations which apart from eating your strawberries, distribute tons of high phosphate fertiliser. The old log heaps covered with blackberry usually have soil incomparably better than the surrounding ground. It is said that nothing grows under this weed. Most native species are light demanding, but if suitable introduced seed sources were present (e.g. sycamore) vast areas covered by brambles would eventually return to mixed forest.

So nature seems set on turning everything into forest – the ideal climax state. So in clearing land we are taking on a battle which we should consider carefully before we bite off more than we can chew. Most of the weed problems in the Australian landscape are due to this last mentioned mistake. More land was cleared than could be effectively managed by the people available. Economic downturns caused people to abandon land for long periods and when they returned (or someone else did) it was head high in blackberries.

Typically, people blamed the plants.

This is all very fine, but what do you do with land in such condition? The answer is bound to be specific to the particular situation.

Nevertheless some possibilities can be considered. But the first question to ask is, "Why do we want to get rid of the weeds?" We need to have good reasons, be serious about working the land in a sustainable as well as productive way before we have the moral right to get rid of the weeds.

Bracken invading pasture land is usually caused by the poor state of the pasture which in turn is frequently caused by continual overgrazing by wild animals, mostly rabbits. As the turf is eaten closer and closer the roots never get into the sub-soil to retrieve the leached nutrients so things go from bad to worse. The bracken unhindered and uneaten by light footed creatures takes over and the soil begins to improve slowly. So shoot the rabbits for food and furs and control your stock grazing by subdivisional fencing (or shepherding if you prefer) and don't complain about the cost or the work.

Bracken Fern

If you have many acres of pure bracken bring in the animal that will eat it – pigs. Rotate them with moveable electric fences and watch them turn your bracken paddock into a ploughed fertilised field on little supplementary feed. Meanwhile, bracken is very fine bedding straw for animals and especially good for poultry deep litter yards.

Or let your vast rugged bracken areas go back to forest, wattles most likely. The silver wattle takes over where the bracken leaves off in improving the soil. It's excellent timber on well-watered soils despite its reputation. Like pine, it rots quickly in the weather, but makes excellent lining boards, furniture, etc., has a lovely grain and figure and is harder and stronger than pine. The period during which it is suitable for felling is limited to a few years before the grubs get at it. Silver wattle is a fast burning fuel, but has a very high heat output. The bark, though not as good as black wattle, makes a tan suitable for skins and furs. Wattle groves are marvellous places to underplant with long lived deciduous and coniferous trees. So if you're into silviculture, silver wattle makes an ideal nurse crop. Its ability to regenerate vigorously on fired ground enables tree cover to be established by simply fencing off pasture, letting the grass grow, spreading seed, or relying on seeds dormant in the soil, and burning. Shelter belts have been established in the Midlands by this method.

Old (30 years) wattle groves make good sites for gardens. Fell the trees and cut them up for firewood. Chop up the tops to encourage decay in autumn then collect the branches and twigs in spring. Burn them and spread the ash. Be prepared to pull seedlings out of the garden for 10 years or so.

Lots of bush country has blackberry growing as a low ground cover. Most of these places were pasture or cleared land, abandoned, recolonised by blackberries and then had fires through, which produced a vigorous regeneration of wattles and Eucalypts and only weak brambles. Removal of the tree canopy will cause the blackberries to spring to life. On areas completely overrun with blackberry, get goats. Cut lines through the brambles to give the goats access. Once well pruned mounds and walls are formed it's easier to harvest the fruit. Get into bees or have a beekeeper run hives off the forage and if blackbirds abound consider them another crop to be harvested – they are not bad baked. Vast areas of blackberries are marvellous sites for hives especially if other seasonal forages are available, as long as some idiot authority doesn't spray them. Try brambles as a nurse for tree establishment and as a guard against browsing stock. Cut holes in the mound and plant out the tree, but keep the rabbits under control with a good dog or ringbarking can kill the trees.

Hopefully this article has thrown more positive light on three plants that many landholders consider a real hassle and will encourage you to think more closely about the role of weeds in nature.

HEDGING YOUR BETS

By ANTHONY WIGENS

You are planning your tree-planting programme for a Permaculture farm. The task of tracking down the species you want to grow lies ahead – meanwhile, your imagination runs wild. You scan the lists of trees considered suitable for temperate regions: honey locust and Siberian pea tree, pecan and butternut, Mexican nut pine and sugar maple. Casimiroa with fruit "like a sweet, green pear", black walnut and Russian mulberry. They seem to offer a more enticing food crop than a harvest of hazel and sweet chestnut.

But think carefully before you go in search of these beguiling plants: the pioneer has the thrill of being a pioneer, but it is those who learn from their mistakes who reap the benefit.

A tree takes a long time to reach maturity and maximum harvest; how many mistakes have you time to make? Certainly we should be introducing as many new varieties of tree as we can; there is nothing new about this, it is the continuation of an established tradition in this country. All I urge is moderation.

No part of a Permaculture plan can be considered in isolation. A tree is not merely a source of timber, fuel or nuts: it is a part of a living system: it provides food and a home for a host of creatures which may not directly affect your livelihood, but indirectly they do.

Diversity is the keynote of organic agriculture. Your planting will inevitably provide a niche for a range of insects, some harmful, some harmless, and some beneficial. If you plant with diversity in mind you are likely to achieve balance – that desirable situation in which the insects that prey on your crops will be kept in bounds by predator insects, and birds, which prey on the pests.

The tree in this country with the greatest number of insect species living in association with it is the oak: 284 species. Next in descending order come the willow (266), birch (229), hawthorn (149) and blackthorn (109). These are indigenous trees: that is, they were native in Britain before 5500 BC, when Britain was cut off from Europe by the sea.

If we look at the trees introduced since that date, starting around the Sixteenth Century, we find that the insect associations are fewer: there has simply been insufficient time for insects to adapt to the introductions as they have to the native trees.

Compare the native hazel (73 insect species) with the sweet chestnut (5 species) which was introduced around AD 100, and the walnut (2 species) which was introduced around 1400, and you can see how long it takes for a new arrival to be accepted as a native by the other residents!

The virtue of the native tree is that the diversity of insects consitute a natural balance. The introduction may survive and prosper, as the sycamore has done (introduced around 1250 and still with only 15 insect species) but the success of sycamore may be the exception, and many foresters see Dutch Elm Disease as a judgement on an exotic introduction. The real English elm is the Wych elm, which has 82 insect species and for a long while proved highly resistant to the disease. The so-called English elm (Ulmus procera) is a relatively recent introduction, which could be why it succumbed so rapidly.

However, the situation is not entirely gloomy. The Forestry Commission publication *Exotic Forest Trees in Great Britain* states "exotics present much the same disease patterns as native trees". There is even a school of thought which maintains that "in the region where a crop was domesticated there are the maximum number of pests and diseases which have been evolved to prey on that particular plant...the farther you get from its centre of origin the more of its pests you can hope to leave behind". This is the argument author Edgar Anderson uses in *Plants, Man and Life* to support the theory that exotic introductions have a better chance of prospering than native plants. He states categorically that "no world crop originated in the area of its modern commercial importance".

This does not necessarily mean, however, that any plant which is removed from its native habitat will thereby prosper. The consequences of such translocation are in general not possible to predict. "The introduction of new crops" says another equally eminent work, "may give native insects greatly improved opportunities of reproduction". (Moreton 1969).

So, try out those exotic specimens, but balance them out with native trees. The best guide I know is "Planting Native Trees and Shrubs" by Beckett.

REFERENCES

Southwood T.R.E. 1961 "The number of species of insect associated with various trees" in *Journal of Animal Ecology* Vol. 30, p 1-8

Anderson Edgar 1954
Plants, Man and Life
Andrew Melrose, London

Moreton B.D. 1969
Beneficial Insects and Mites. MAFF.

Beckett K & G
Planting Trees and Shrubs Jarrold

IMPRESSIONS OF N.Z.

By DAVID HOLMEGREN, April 1979

On this trip to New Zealand, my first, I have focused my attention on the drier hill and high country of the South Island. My first real view of the country was in the Waihopai-Avon area of inland Marlborough.

Coming up the Wairau valley the lack of trees on the vast rugged landscape was obvious. The pine shelter belts with their tops bent by the prevailing winds told of the icy winds which must chill people, plants and animals. The rivers shocked me; wide shingle beds where there should be productive flats. Steep, thinly grassed slopes, barren tops and scree, wildly fluctuating rivers causing untold damage, all indicate landscapes in poor shape. The country as a whole is visibly young and unstable, eroding and decaying very quickly. These are natural consequences of a geologically active land and to most people seem inevitable.

VITALITY

The other aspect of the country which took me a little longer to see was the vitality of the life forms, native and introduced. Life, primarily plant life, is the way the chaos of upheaval, volcanism and erosion is balanced. Order, stability and conservation are the characteristics of living systems; ie ecosystems. Tremendous vigour in the plant systems is needed to colonise and conserve the wildly changing earth. The paradox that struck me is that the vitality of the life here comes directly from the chaotic forces of land movement which provide a diversity of micro-climates, drainage and soils. The young soils in this country, in spite of their poorly developed ecology, have the capacity to support vigorous plant growth.

In parts of the Avon Valley I have seen the regeneration of native forests which was most encouraging. The pioneer 'weed' Kanuka covering the shady slopes provides conditions for the Lancewood, Mahoe, Natipo, Olearia, Coprosma, Wineberry, Tuschia, Broadleaf and Beech to grow and thrive. In places the process was more advanced with Kanuka (many 100 years old) remnant in vigorous beech forest. Totara and Matai specimens indicated the ancient Podocarp forest which once grew there.

The streams in these forests are stable with bogs and flushes marking where minor tributaries enter, acting like massive sponges storing water to be released over dry periods. Large numbers of wasps and bees working the honeydew and other forages indicate a potential yield. Skins and hides from feral animals are another product. The yields should be considered as bonuses because the real value of the steepland forests is to provide stable aquifers and river systems which are necessary for intensive long term management of the flat lands where most of the people live. Japan learnt centuries ago that to maximise the stability and long term productivity of the flats, steeplands need to be in forest. New Zealand must learn the same lesson.

INTRODUCED PLANTS

Before coming here I was aware of how a huge range of introduced animals flourished but I had no idea of the vigour of introduced plants. In Tasmania, which has a similar climate to much of the South Island, many plants which are naturalised here have to be nurtured as garden plants. We do have some introduced 'weed' shrubs and trees – Gorse, Sweet Brier, False Tree Lucerne, Willow and most widespread of all, Blackberry. Here I have seen Pines, Spruce, Larch, Fir, Alder, Ash, Poplar, Oak, Hawthorn, Barberry, Rowan, Cherry, Plum, Gooseberry, and many others naturalised, and the landholders are struggling to exterminate them by the use of dangerous chemicals. The 'weeds' will win.

For me it was interesting to see native Tasmanian Eucalypts and Wattles very healthy and naturalising, but not the Blue Gum *(E. globulus)* which has been so widely planted in the Canterbury region.

The dieback of previously healthy trees, reminicent of bushfire damage in Australia, is due to massive attack by Tortoise beetles *(Paropsis spp)*. The lack of predatory and other natural controls which exist in Australia are probably responsible for the severity of the attacks and indicates the importance of planting a wide range of species rather than one "proven" type.

An exceptional example of Eucalypts and Wattles thriving was at Hawkeswood in North Canterbury. What appeared from Highway 1 to be a typical

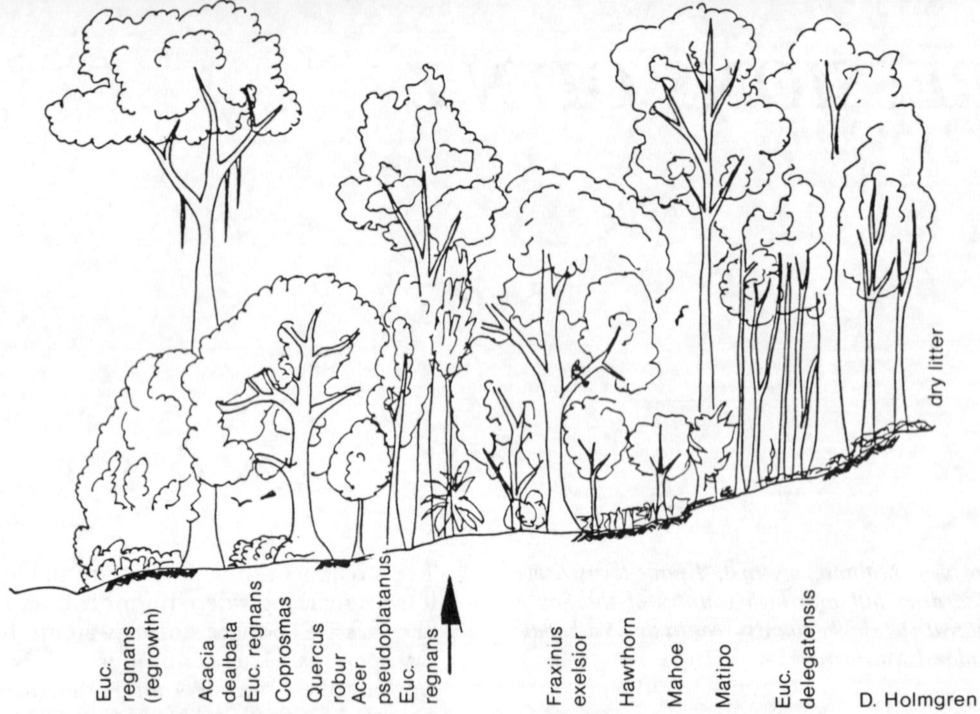

Tasmanian wet sclerophyll forest was in fact a whole new forest system in the making. It shows potential that no pine plantation could. Although I have little knowledge of its history, I will go into some detail on its present state and possible evolution to show the permaculture way of looking at the land.

FROM A PERMACULTURE PERSPECTIVE

The dominant trees are Eucalypts between 100 and 150 feet tall, the larger ones being *E. delegatensis and E. regnans.* Elm, Oak and Ash up to 100 feet look to be of a similar age. The tops of the tallest trees were windblown from the prevailing wind, suggesting they had been planted out on open ground. All the larger trees showed signs of fires and there were some burnt and dead trunks of deciduous trees. Two distinct generations of Eucalypt regrowth, obviously from the fires, had regenerated thickly but not under the large deciduous trees where light levels would have been too low. The older fire regenerated Eucalypts were up to 120 feet, very straight and only 2 feet diameter, indicating an age of less than 50 years. Silver Wattle *(Acacia dealbata)* a short lived, fire regenerated nitrogen fixer, was abundant and formed a scrubby edge along the highway. Blackwood *(Acacia melanoxylon)* was regenerating in a gully from an old straggly 50 foot specimen. The younger ones look destined to be tall straight timber grade trees. Young Seedling of Oak, Elm and Ash were coming up through an understory of Sycamore, Blackberry, Plum , Hawthorn, Gorse, Mahoe, Coprosma, Black Matipo, and Cabbage Tree. Sycamore and most of the natives seemed to be taking over as the dominant understory from the thorny 'weedy' shrubs.

As would be expected in a wet sclerophyll forest the Wattles and Eucalypts, except Blackwood, were not regenerating and unless another fire came through, the Eucalypts are destined to be succeeded by deciduous trees, and native trees and shrubs. Each plant paves the way for others which will succeed it leading to longer lived species and a more stable forest type.

SUCCESSION

This process of succession could be directed by suitable underplanting, to provide increased yields and continue the soil and aquifer development already occurring. According to a N.Z. Forest Service hunter familiar with the area, enormous numbers of pigs have been culled from Hawkswood, indicating the great forage potential. The value of selectively logged trees was obvious and some thinning has already taken place.

Tentatively, I would like to suggest underplanting with Black Walnuts, European Beech, Horse Chestnut, Hickories, and some of the American Oaks. These are all long lived, relatively slow growing trees of special timber value which also provide useful forage (nuts and seeds) for pigs and deer. These animals could be farmed allowing for resting and rotations to enhance the natural regeneration.

Honey and structural timber particularly from the Eucalypts and Wattles would be a continuing yield for some time, but future timbers and animal products would be the main yields in the long term.

I hope this small account of Hawkswood explains how permaculture principles might be applied in the N.Z. landscape. Permaculture is not just forest farming. It is the way of working with nature to provide livelihoods for ourselves and our descendants.

COMMENT

David is one of the best observers of landscape I know. His reading of nature should encourage us all to open our eyes to our environment.

CHANGING PATTERNS OF BEE PASTURAGE IN NEW ZEALAND

By DICK WALLINGFORD

Bees are necessary to agriculture as pollinators. As the bees' traditional forage sources are eliminated, areas that still provide nectar and pollen from well-chosen plantings being made today will still have a viable honey production potential. Other areas may need to depend upon hives hired for pollination service to ensure adequate pollinator numbers.

Tendencies of modern agricultural methods in many areas have been to the disadvantage of the beekeeper. At one time bees could gather a considerable crop from roadside flora and from weeds growing within established crops and pastures. Dandelion *(Taraxacum officinale)* has traditionally provided nectar and pollen over a period of time in the spring when it is especially needed in the hive. Though it is not a high grade pollen, lacking one essential amino acid, it helps in the colony's spring build-up as long as other pollens are available to complement it. With improved selective herbicides and different methods however, this important source and others such as buttercup (Family Ranunculaceae) have been cut back in many localities.

Another plant being eliminated because of its noxious qualities is gorse *(Ulex europaeus).* As old established hedgerows of gorse give way to high tensile wire on the Canterbury plains, beekeepers there have encountered pollen shortages of a magnitude to inhibit their hives' abilities to rear healthy young bees in the spring. They have been forced at times to either shift hives nearer to the river beds where diverse sources are still available or else attempt to supply pollen substitutes to the bees. Either of these involves expense to the beekeeper and will be reflected in the price of honey. Though Canterbury is well suited to produce large crops of desirable white clover *(Trifolium repens)* honey, this early spring pollen shortage makes beekeeping there somewhat unpredictable.

Blackberry *(Rubus spp.)* and thistles *(Cirsium spp.)* are other noxious weeds which provide forage for bees. Their suppression has contributed to smaller honey crops, only offset by increased efficiency in management and transportation making hive-shifting economically feasible.

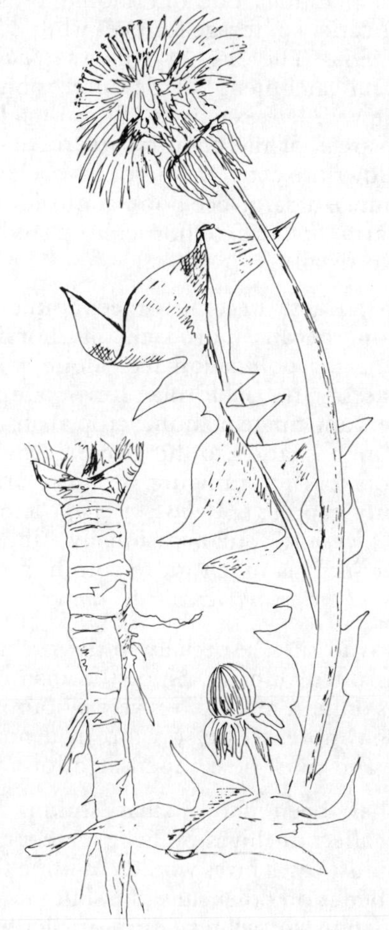

Taraxacum officinale

Bees are an integral and necessary part of agriculture and horticulture, and beekeeping in some form must remain a viable occupation. Feral, or wild colonies of honey bees are not sufficient to support the increasingly intensive pollination demands of modern agriculture, and the expansion into the presently marginal lands will further reduce their numbers. *To maintain or increase the number of hives available for pollination of crops and pastures there are two main directions that beekeeping in the future will take.*

1. If beekeeping for honey production remains

economic, bee populations will generally reflect an area's ability to supply pollen and nectar. As mentioned above, however, the "clean" nature of modern management has and will seriously limit honey production for profit in some areas of the country unless some action is taken.

The choice of plantings today will greatly influence the nectar and pollen available to the hives in the future. Shelter trees, amenity plantings, and other significant plantings can and should be selected to provide a variety of forage for bees while still meeting other criteria. The use of willows *(Salix spp.)* by farmers and catchment boards for erosion control is one of the very few examples of this on a large scale. In many areas of the country, riverbeds lined with these willows are actively sought by beekeepers, and this assurance of large bee populations for the nearby areas of farm, orchard, or horticulture land should be a lesson to us all.

2. When an area becomes uneconomic for honey production because of lack of floral sources, beekeeping for pollination fees alone will result in many cases. In California for example, many beekeepers anticipate no honeycrop at all, depending entirely upon charges to the grower. Charging such pollination fees for bringing beehives onto a crop, particularly apples *(Malus pumila)* and kiwifruit *(Actinidia chinensis)* but occasionally others such as clover for seed, is much more common than in the past. *Several factors influence these charges:*

(a) Access to sites, especially in the orchards in the spring, is often difficult. Since the majority of hive shifting is done at night and is very labour intensive, it is just one more activity on top of a normally full schedule and must be compensated for accordingly.

(b) It has been shown that shifting hives can adversely affect the hives' ability to gather a later crop. Since in most cases hives would continue their spring buildup better on other sites, the shift into and out of the pollination site will result in a smaller harvest later in the season. Though hives will gather nectar from many of the crops they pollinate, in general the hives could have been better brought up in strength on other sites.

A bee hive constructed in a bagged bunch of bananas found on a banana plantation.

(c) Pesticide usage can be a constant worry to a beekeeper, even when a grower is aware of the risks, and plans his spraying programme with the bees' welfare as well as his own crops' in mind. Since bees may forage up to several miles from the hives, indiscriminate or inadvisable pesticide application on neighbouring crops can pose a threat even with the best of precautions. Approximately twenty per cent of the colonies in California are lost to agricultural chemical usage each year.

The presence of bees as pollination agents will enhance the crop returns many times over with certain crops. Charges to bring hives onto the site are recovered by the grower through this increased yield, and beekeepers are repaid for their labour and loss of production. If adequate foresight in providing future nectar and pollen sources is used now, such charges will remain modest, but if in the future a beekeeper must rely to a greater extent upon this revenue due to decreasing honey crops, the charges will most likely increase.

URBAN FORESTRY

By JOHN FRENCH

Urban residents in Australia have a love for trees and shrubbery easily seen in the large amount of vegetation in the average suburban block, and the protests that often follow when trees are to be removed to widen a road. New housing developments often gain a "forest" appearance when viewed at roof top height 10 or 15 years later. Unfortunately, existing land subdivisional procedures and street arrangements restrict the scope of the residents in planning "mini-forests" and no opportunity is given to develop a neighbourhood forest environment which may be of great benefit to the community. Local Government Councils plant "linear forests" on nature strips along streets and plant trees in urban parks.

Urban forestry is a specialised branch of forestry. It is a method of intensive land use. Trees are planted, maintained and harvested in cities – not only for their shade and beauty, but also for their commercial value. These forests not only give city dwellers wood products and recreational areas, but also act as wildlife sanctuaries.

Over 86% of Australians live in urban situations, and Australia is one of the world's most highly industrialised and urbanised nations. Over 11000 hectares of prime arable land are submerged annually by expanding cities yet there is no counterbalance by which an equivalent area of land is made more productive. About 30-40 percent of this urbanised land is paved, with impervious materials, which accounts for nearly 3,800 hectares of land permanently removed each year from any form of production. We can't compensate for this indefinitely by pushing farm lands further out. Apart from the warnings of the current energy crisis, and dwindling phosphate availability, there is obviously a limit to this sort of land use.

In Australia there are approximately 36 million hectares of forested land, of which only 16 million hectares, can be classified as suitable for production of timber. Projections of population growth and economic activity suggest we are heading for a timber famine. As a result there is a contrived move by some to increase exotic pine plantations to meet this demand. Urban forestry proposes that we create a new kind of forest resource – the urban forest. Such urban forests would help meet demand for timber products, lessen the pressures on natural forests, and enhance the quality of life for urban dwellers.

In Europe, foresters have managed city forests for many years. The timber cut from these forests not only helps to pay for the maintenance of the forests, but helps pay for other city costs as well. Most are not owned by governments but by communities. In Switzerland, for instance, 70% of the forest are community forests. 24% of Zurich is forested. Urban forestry is established on a small scale in Canada and the U.S. and a great deal of research done regarding tree species and their adaptability to urban environments.

Although there is no planning at a national level for urban forestry in Australia, some important aspects of it such as "greenbelts" are incorporated into the layout of Canberra, and are proposed in the designs for Albury/Wodonga (NSW/Vic) and Monarto (S.A.).

We need to plan the expansion of Australian cities, so that we not only protect those resources which should not be built upon e.g. highly arable land and historic and religious sites, but also prepare the environment for its future population. We must plan and set aside forested areas for times to come.

Because we have no formal training in urban forestry in Australia and most forestry operations are carried out in country areas and deal mainly with the management of trees on thousands of hectares, town planners, foresters, horticulturists, sociologists and other natural scientists should co-operate as part of a "multi-disciplinary team", to develop a land management policy nationally. However, it is important to remember that in all countries where urban forests are established, there is a great deal of evidence to show that community involvement in the development, maintenance and care of these forests is essential for their continued and efficient use.

The key is education at all levels of the community, from schools to businesses. Community involvement in planning and operation functions should be encouraged, making people aware of the investment costs, obligations and responsibilities of their forests as well as the benefits they stand to gain.

Trees can be used to cleanse sewerage and add considerable amounts of water to the ground reservoirs.

Furthermore, trees add to our aesthetic values, and Real Estate Agents are well aware that trees increase property values.

There are suggestions trees have an unconscious recreational value. Many people can relax just by being among trees; they provide us with quiet, and recovery is enhanced by such surrounds.

We could create wildlife habitats. Urban areas could be systematically stocked not only with desirable species of trees, but with desirable species of animals, birds and insects.

Urban forests also affect the environment in other ways. Trees provide valuable shade and lessen dust fall, air pollution (trees clean the air too), wind and noise. Trees soften the impact of man-made materials such as bitumen, concrete and steel, and produce an environment more intimate and private. Trees also cool and filter the humidity of air. Local climate would be changed favourably, cooling city centres in summer. A drop of 1 degree C reduces costs required to cool buildings by 25%. In winter, trees help reduce heat losses at night by reducing heat radiation from ground surfaces. If trees now pruned or destroyed by local councils, CRB and SEC, were made available 'to the public' as fire fuel or compost, much wastage of valuable resources would be avoided.

A natural drainage system could be set up, using porous pavements, the trees and vegetation, soakage pits for roof run off and grassed open drains as catchment and drainage facilities, reducing engineering and maintenance costs.

In the Woodlands Project (Houston, USA) a natural drainage system was demonstrated to cost 22% less than a conventional system. As well as increasing soil aeration and water catchment.

Appropriate operations should be handled by residents, school children, service groups and clubs, under the supervision of trained Foresters. Urban forestry involves all segments of society and would help to develop a "land ethic" in people and foster a love of the outdoors and forest land. They would become an "open classroom" for all people, so that environmental education would be part of our general education.

Obviously in the already densely populated areas of our cities, we could not expect to establish vast urban forests, but we could provide more trees and more species could be planted and cultivated for timber along nature strips, highways and adjacent to trainways, and on recreational areas. Many city streets in the older suburbs could easily have 3-4 rows of native or exotic trees grown along their centre length. Urban "greenspace" will be recycled space – reclaimed dumps, old building sites, industrial wastelands. Urban forests could be of varying size, from .5 ha. to 200 ha. The Urban Forester has a large choice of tree species to choose from for the forests. Large, medium and small sized natives, interspersed with shrubs, fruit and nut trees, shade trees and honey trees etc. could all be grown to suit a variety of needs and uses.

ROOF GARDEN IN THE SMOKE

By ANNE-MARIE

We were advertised to start at 10:30 and we wondered if anyone would turn up. 'We' were Sue Carpenter whose flat it was, and who had planned and organised the event, Bob from 'Green Deserts' and me, Anne-Marie. And the event itself? To create a mulch garden on the roof of the block of flats that Sue and some 20 or 30 other people had their homes in.

All the necessary materials had been collected by Sue over the preceding weeks – straw and horse manure from the nearby Kentish Town city farm, sacks of food waste from Sue's own kitchen, also from the Camden market, the Castle Pub, and a local French restaurant; there was also plastic sheeting, from the carpet shop up the High Street, to line the roof and prevent leakages.

Amongst the pile of rubbish which had been dumped alongside our materials by residents who had mistaken it for a new communal refuse tip there were some very useful containers – an old fibreglass car bonnet, an Aladdin oil lamp and some old metal filing cabinets, a flock filled bed base and old horse-hair matress for soil substitutes, decayed rubble for lime, and some old carpeting. The only thing missing was earthworms and these were thoughtfully provided by one of our five co-workers who arrived during the course of the day.

Sue Carpenter and John Knights, on site.

The first job was to clear the space of rubbish and sweep it as clean as possible. There was quite a bit of discussion as to how to line the roof, and finally we overlapped the strips of plastic to follow the down-flow.

We took great care to line the parapet walls too since we did not want to cause any rot to the fabric of the building. This was a strong roof, but great care should be taken if planning such a venture that you choose a building whose structure can take the additional weight.

The old bed base made a perfect container for the first mulch garden. We turned it upside down, ripped off the sacking cover and started by adding horse hair to the flocking that partially filled it. We decided to use this for a soil substitute to hold the decomposing food waste and stable sweepings which would provide the main body of the mulch. Since both of these are organic they would rot down too in time and provide added nutrients especially nitrogen. We thought that if the final results proved too rich, a couple of sacks of sand added to the bed might improve it, but in the meantime we added some old rubble and mortar for lime.

The worst part of the job was adding the food waste, for Sue did not know that if the waste were left in plastic sacks the initial decay would be anaerobic. The stench as we opened them knocked us back, but we got over our squeamishness in one way or another and spread the sometimes glutinous mess on to the base material.

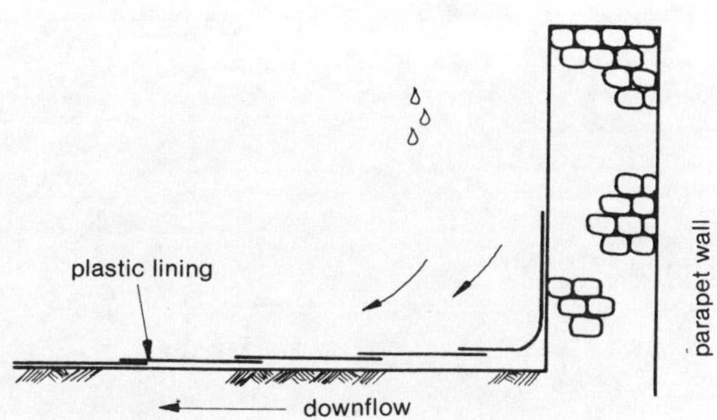

Finally we topped the whole lot with stable sweepings from the city farm. This gave a very pleasing appearance and we were quite excited by the finished result.

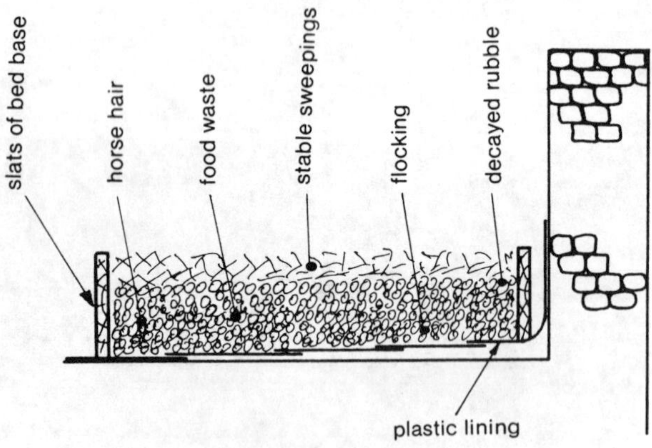

Once we got the format, we quickly devised several more beds and containers filled on the same principle, though where the containers were very deep we filled them first with any old rubbish that we wanted to get rid of.

By the time we finished there was little left but a neatly folded pile of plastic, a little wood, some carpet, and a last sack of stable sweepings to start the next bed.

They looked good and we wanted to plant them straight away. So after a leisurely pint we went to the newly opened Camden Garden Centre. There we bought seeds, plants and potting compost to start them off in.

We spent some time discussing how to plant up the seeds, and in the end the problem was solved by Peter, the ex-tobacco farmer from Zimbabwe who had seen Sue's poster in the newsagent's. He had found a piece of polystyrene packaging that used to hold yoghurt pots or something similar, ideal for forming pockets of newspaper to be filled with soil and seeds, then to be extracted and planted in the mulch. We could do this indoors.

Sean made drills in the mulch beds too by drawing back the mulch, trickling soil along the furrow, adding seed and a topping of soil, and finally drawing a thin layer of mulch over the top to deter the birds. We added two ready-grown plants to complete the picture and leave any visitor in no doubt as to the purpose of our construction.

We had carted several buckets of water on to the roof when the mulches had been completed but now we were helped by the timely intervention of a downpour to complete the process.

And there it was, Sue's roof garden: source of much fun, discussion and laughter in the creation, with those few stifled retches as we opened the rotting sacks of garbage. To be turned we hope into the sweet smell of success, in the form of future produce, and a little fresh green space in the heart of this mainly concrete and asphalt city desert.

COMMENT:

A roof garden and self-sufficiency? Not quite, but a very useful addition of greens and herbs to every meal.

I guess the story is so exciting to me because I was at the birth of the idea. It was on the way back from Blencarn (Cumbria) when Sue invited a few of us to have a look at "her roof". About six or so storeys above the grey pavement of London. The corner of the roof offered a warm micro climate, high up where the air is cleaner.

And now it's a garden! Fantastic! Who needs 5 acres?

Max

MARYBOROUGH'S PERMACULTURE PONDS

Maryborough's (Vic) innovative Council asked Bill to design them a Permaculture for their sewage settling ponds. This, slightly abridged, was his reply.

By BILL MOLLISON.

(Photos supplied by Terry White)

OUTFALL LAGOONS

These contain tertiary-treated water of good quality and present both opportunities and problems. The problems are those of excess nitrates / nitrites in released water, causing eutrophication of the Loddon-Murray area and local streams. The opportunities are to solve such problems in a constructive way by utilization. The aim is to develop a productive system releasing clean water to the Loddon catchment.

UTILIZATION WITHIN LAGOONS

(1) A considerable wildfowl population (duck, coot, terns, waders) use the lagoons, helping control insect and weed populations. Subject to predation by feral cats and foxes, the ducks in particular have no refuge area. It is therefore recommended that in any future lagoon evolution, islands are provided, planted with bamboo and willow, and provided with hollow logs and nest-boxes for a wildfowl refuge and breeding area.

(2) Vegetation is already establishing along lagoon walls, but this should be a planned development of useful aquatic emergent species and shrubs yielding some crop for craft or wildfowl food. Sweetrush, tree lucerne, lucerne itself, comfrey (as root sets), seeding bamboos, *(B. macrosperma)* and like species would suit. This planting also, by transpiration and nutrient uptake, is beneficial to water quality. I have indicated shallows around new lagoon corners as a substrate for water lily and like plants, with either a commercial sales potential or a functional (shelter and food) potential for wildlife.

(3) Aquatic animal species such as mussels (available locally) and some local fish species can be introduced in trial area (fish perhaps in floating baskets to test water quality). Small fish control mosquito populations, and some larger species (grass carp) provide control for algae and weed.

(4) Future lagoons should be subdivided into 100ft. wide sections to permit the efficient harvesting of fish species. Fish so reared can be transferred to clean ponds for a week or so before sale.

A complete bird, plant, animal system will make a great deal of use of nutrients, and 'spin off' species and products, plus free-flying duck for the state.

By building low bunds and installing flood valves from lagoons to planting areas, a good deal of water can be absorbed in crop production over the 50 acres owned by council. Bunds, flooded, can produce mints (for oil production), celery, asparagus, and like food crop (to be treated or cleansed before sale), and lucerne, comfrey etc. for forage crops for sale.

On the bunds, bamboos and small fruit can be grown for harvest, as well as hazelnuts, citrus, and other low crops.

A main bund and drain should parallel the crops, and on this, larger crops (pecan, walnut) planted. Inside the main bund a main drain should receive excess water from the paddock or small bund systems via small sliding floodgates.

Water from the lagoons can also be conveyed to much larger areas as nutrient spray for sunflower, lucerne or other forage and oil crop, either on areas owned by the council or leased to local landowners, or charged for in private use on golf course and like areas. At about a quarter million gallons per day, large areas of up to 1500 acres can be treated in this way.

The result is that nutrients are converted to useful crop, water is filtered by transpiration or soil penetration, and only clean water escapes to air and streams. Great productivity under lease or council direction can be achieved for annual crop and perennial forage. Food crop and bamboos or canes form the basis of small local industry, together with oil distillation or pressing from oil crop.

Parks and garden staff at Ballarat or Melbourne Botanical Gardens can obtain and suggest various useful aquatic and bund species for reproductive use. Areas can be leased to garden society, growers, youth (unemployed) or craft groups and harvest of the crop also leased.

Such plantings will also aid the problems of salting in the Kerang area downstream of Maryborough.

As well, quite complex gardens could be developed by lessees on smaller sites.

Sewage sludge or mud from lagoons, via tanks can make establishment of vegetation on old mined-out areas much easier, as can small local silt dams of stone, logs, to trap silt runoff, and these sites planted to carob, honey locust, mesquite, and olive as a town resource.

Maryborough can see itself as a pioneer city in the Loddon area and would become a mecca for visitors if such schemes were started.

Techniques of sewage disposal and fire control are

Bunds with newly planted albacutya red gums.

Sliding floodgate.

needed by many towns, and both citizens and councils would benefit from a tour of any areas you developed.

SPECIES SUGGESTED

In Lagoons	*In Bunds*	*Small Bunds*
Arrowhead	Asparagus	Strawberry
Common Reed	Artichoke	Bamboo
Waterlily	Comfrey	Cape Gooseberry
	Cranberry	Gooseberry
	Lucerne	Hazlenut
	Mint	Lemon
	Rhubarb	

Main Bund	*Spray Irrigation*	*Area Islands*
Feijoa	Lespedeza	Willow
Hazelnut	Lucerne	Pampas
Lemon	Comfrey	Bamboo
Loquat	Sunflower	
Mulberry		
Pear		
Pecan		
Chestnut		
Walnut		

REPORT

Mr. W.J. Mackay, City Engineer employed by the city of Maryborough, offers these comments on the Permaculture initiative at the sewerage ponds;

"Bill Mollison visited Maryborough several times in 1977 –78 and gave a report to council, suggesting uses to which the effluent could be put.

Maryborough, which is a town of 8500 people, discharges an average of 600,000 gallons of waste water per day, (2730 cubic meters), into a series of oxidation ponds known as the waste-water farm. This effluent is previously given primary and secondary treatment through digesters and trickling filters.

Bill Mollison gave a report to the council setting out recommendations for development of the area.

As a result of his report, several islands have been built within the four lagoons, and this has helped the

Terry White inspects a planting of mixed eucalypts and casuarina.

bird population. Sadly, some of the tree lucerne, bamboo and other species which were planted around the lagoon edges, have been eaten by cattle. However, pampas grass and some shrubs have been planted around the edge of the lagoon which now look most attractive and the banks are firm with a large amount of vegetation.

Redfin and Yabbies have been introduced to the lagoons and are no doubt breeding at the moment.

A large number of native trees have also been planted along one side of the area and the overall appearance of the farm area is vastly improved from the sterile area which existed before Bill Mollison came on the scene.

In conclusion, I suggest that the ideas envisaged would be of great benefit if proceeded with and it is hoped that in the foreseeable future, this work will be carried out."

WHAT IS A HAMLET....THAT IS THE QUESTION

By DAVID CAMPLIN

There has been a lot of talk about Hamlet Development as a practical zoning for the so-called alternative lifestyle but few people agree as to what the word means. I personally would be the last one to try to put a label on, or a box around, any concept for I believe that the freedom of an idea lies in continuing growth and life.

However, the necessity of order as understood by a practical town planner, requires some definition. So what is a Hamlet. The Oxford Dictionary defines a Hamlet as a small village without a church. Not that the people living in a hamlet necessarily lack faith, it was just that they usually lacked the population and finance to support a permanent home and a qualified teacher or preacher of the subject. Which may explain why some of the greatest prophets and prelates were born in and grew up in the simplest and poorest of rural hamlets. But this is one of the side effects of Hamlets as a human environment. For the very simplicity of its inhabitants, life dictates each individuals' philosophy and actions which together become the community's life style.

The precendents of the past also show that traditionally hamlets shared other factors in common. They were confined to rural areas. Whilst we may consider this fact self evident, concerned public servants may not and I would be horrified to have them rushing around making up reasons and by-laws to prevent hamlet development in areas zoned "light industrial" or "residential" or whatever. So they may seize this first clue as to what a hamlet is and turn accordingly to the vast expanse of the shire map marked "rural".

They were predominantly confined to marginal agricultural land. As the original pioneers of past hamlets were rarely blessed with an abundance of material wealth, hamlet buildings rarely, or ever, intruded on prime agricultural or pastoral land. This is true even today where, to my knowledge not one of the quasi legal, hamlet type communities that are springing up like mushrooms have been established on or destroyed the potential of a profitable farm.

In my opinion, hamlet development would actually complement the mechanized mono-culture type

David Camplin. Hamlet proponent, poet and writer.

agribusiness of today; providing a buffer zone between potential and proportionate pest and plague areas. This fact should assuage the anxiety of concerned food processing corporations such as milk boards, sugar mills and canning factories who could otherwise see hamlet development as a threat to their production and profits.

Hamlet buildings and facilities were low cost self-help developments. Strangely enough the shoestring budget buildings of the traditional European hamlet have an unique charm and beauty with their creative use of locally available materials and designs. These aesthetically beautiful functional forms, evolved through necessity. Low cost stone crofters cottages, thatched farm houses and log cabins all have a charm of their own. And as they were usually ownerbuilt, with love, to live in, they have often outlasted the castles and manor houses which could not adapt to the winds of change.

68

As hamlets were based on the need of self help, owner building and occupation, the abuse of this form of development by speculative real estate subdivision, spec buildings and absentee landlords could be avoided. Particularly so when both historically and practically, the accustomed modern services of power, water, sewerage, drainage etc., is the responsibility of the individuals concerned. The existing rural parkland zoning requires that these services be available prior to residence, and the very cost of these services precludes any possibility of ownership apart from the privileged rich.

Hamlets have naturally been ecologically and environmentally orientated. As occupants of hamlets are naturally and automatically involved with seasonal changes and the balance of nature, it is rare that they abuse this balance with the wilfull destruction of the native flora and the subsequent reduction or total elimination of the native fauna. Hamlets normally grow as part of the natural environment, access normally follows natural contours, drainage follows natural watercourses, buildings are orientated according to nature and human harmony with nature evolves. If it doesn't, the power of nature reasserts itself and the hamlet is destroyed through the natural forces of wind, water, fire or famine.

Hamlets develop as economically viable semi-self-sufficient satellites of larger market villages and towns. Historically hamlets weren't planned but arose as a result of human need. Once established they developed a degree of agricultural self-sufficiency, predominantly food for local consumption and supplemented minor cash crops production with the manufacture and sales of cottage industry art and craft products, and supply of casual labour.

To my knowledge there is no allowance within zoning regulations for the arts or crafts of cottage industry. I think that by definition I am breaking the law by writing this article at home…my only defence being that as it is for no material reward it may not be classified as a profession. However once a sale is achieved I would theoretically require an office to continue.

The same principle holds true if you make jam to sell on a street stall, or crochet, weave, tatt; make sandals or candles, pot (as in pottery) or design posters for profit. Those who complain of a lack of culture at a grass root level should first examine the environment in which this culture can survive. And there is nothing more stifling than the sticky clay of zombie zoning to rot the roots of even the hardiest of native plants.

Town planning has given us the inhumanity of Canberra, the vandalism of Green Valley, and the waste that was Monarto.

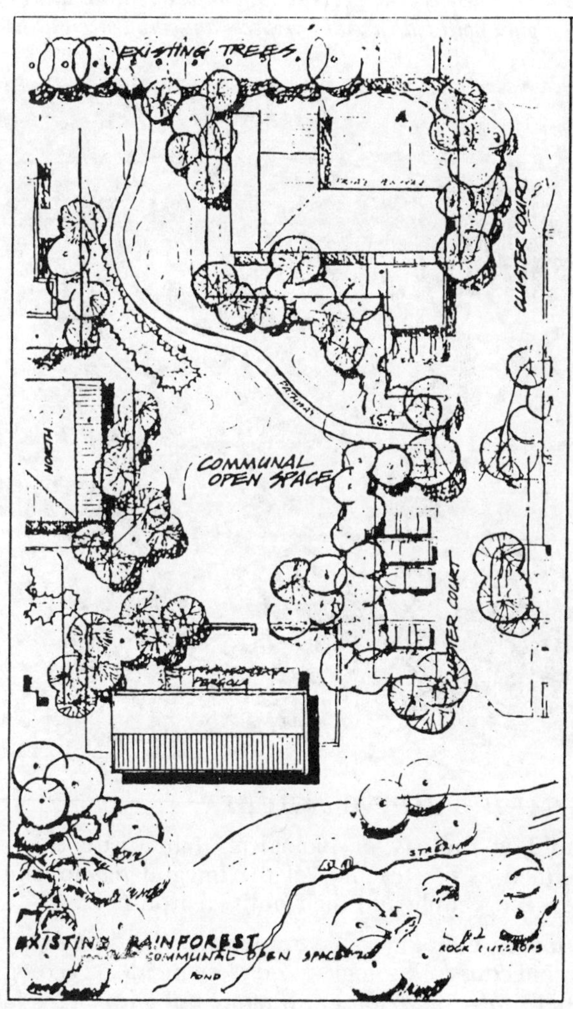

The modern cluster housing concept. It contrasts the uniformity of conventional subdivision, retaining natural beauty and is developed in consideration of the local environment.

I can only pray that the people responsible and who have the authority for shaping our living environment, can somehow broaden their thinking sufficiently to accept the principles of natural growth and somehow include these concepts, which I have tried to do in defining a Hamlet, back into the framework of our Society.

THIRD SKIN
BUILDING BIOLOGY

By ALESSANDRO VASELLA,

In present discussions of ecological problems we seldom talk about the environment within our homes, where we spend over 90% of our time. There is evidence that the four walls surrounding us most the time — especially in modern buildings of recent construction — are highly polluted in many respects and influence our health in a negative way.

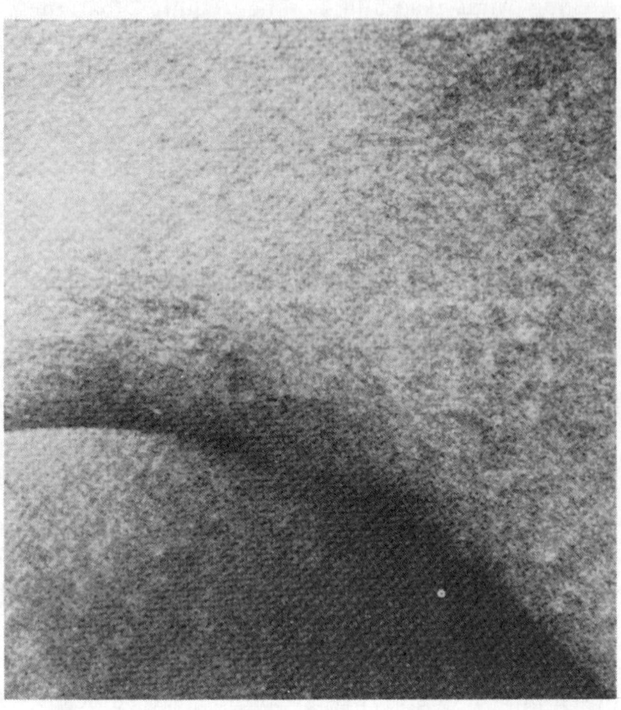

(Photo, A. Vasella)

WHAT IS BUILDING BIOLOGY?

Building Biology or biological architecture can be defined as the teaching of the integral relationships between people and their built environment.

Building biology works on many levels. It deals with architecture, biology and medicine, ecology, philosophy, psychology, to name but a few. This fact makes the recognition of Building Biology as a "new science" difficult (as is the case with Permaculture) because it cannot be put into one category. Building Biology aims at a healthy living and working environment by preventive measures. According to the World Health Organisation health is a condition of complete physical, psychological and social well being.

Technology is not aiming at inner quality but in most cases at measurable quantity only. Building Biology tries to restore or re-establish the lost balance between technology, culture and biology. The three should play a role of equal importance in the building activity.

Looking at man in his whole evolution he still is a hunter-gatherer of the pre-agricultural period and has in fact not adapted himself to technology (See Rifkin, 1980). We should try to adapt technology to the biological needs of man and nature in general and not nature to technology!

THE THIRD SKIN

In Building Biology we consider the building as an organism. To help this image the walls surrounding us can then be seen as the skin of which we expect many essential functions. We simply talk about the *third skin* (our clothes being the second skin).

Our skin is a very important organ. Some of its essential functions are: breathing, evaporating, absorbing, protecting, insulating and regulating. There is a constant exchange between inside and outside and we talk about open systems.

If we translate these skin functions to our shelter we find that certain things we do in modern buildings are quite wrong and that we have transformed many buildings into almost closed systems. Take for

70

instance the insulation euphoria of recent years which has had us all packing up our buildings with insulating materials which usually have a damp proof course on them. By doing so we especially fail to take advantage of the passive solar energy during the long spring and autumn periods of temperate climates. For instance, the sun heats up a brick wall during the day and this stored energy is transported into the building during the night and thereby prevents the rooms from cooling too fast. In an insulated building the walls have lost their heat storage possibility and in

spite of the insulation the energy saved is quite less than that lost. In addition we seal off all the cracks of windows and doors. Finally a building 'treated' in such a way cannot 'breathe' anymore.

But constant air renewal is essential for a comfortable and healthy climate. This is a severe problem in such buildings and may even result in an accumulation of wastes and poisons in the room air. Air conditioning needs to be mentioned in this context. It solves only the quantitative part of the problem. We can remove dust particles by filters, add water for the necessary humidity and heat or cool the air. But by the time the conditioned air reaches the rooms its quality is lost. The friction in the air ducts creats an imbalance between positive and negative Oxygen ions (the optimal ratio is 40:60). In addition the 'conditioning' of the air results in the accumulation of large ions.

MODERN BUILDING AFFECTS ENVIRONMENT AND HEALTH

Many former natural materials from renewable resources have been substituted or denatured by artificial or synthetic products. These processes often make these materials unrecyclable. For instance the real fire danger in modern buildings is the toxic gas from synthetics (paints, varnishes, textiles, furniture) and not so much the fire itself. Usually a new material gets approved by an official material control laboratory. The problem is that the testing is restricted to the physical (measurable) or quantitative part of the material like strength, heat resistance, insulation capacity, and so on. What is rarely analysed is the influence of such a material on the health of the inhabitants of a house. I mean the biological quality (e.g. its radioactivity, release of toxic gases, or the filtering capacity).

The raw components of a building material are a vital factor for its biological quality. Big differences of radioactivity for instance are found in cement, bricks and plaster. Usually the radioactivity is high if the components come from a blast-furnace (slag) or from aluminium production (red mud). All the synthetic resins as they are used in chipboard etc. leak formaldehyde (a very toxic gas) in small quantities but over a long stretch of time. The degrading part of

many plastic materials is invisible but often toxic. That's why many plastics get brittle with age because they have lost a component usually in the form of an invisible but undesirable gas.

WHAT ARE THE HIDDEN COSTS?

The modern building industry is unecological in many other ways. The irreplaceable top soil is frequently lost and buildings sited on prime productive land. The removal of soil from the site for lower stories, the pollutive and energy consumptive transport of materials over long distances (e.g. in the pre-fabrication business), and a high energy input from non-renewable resources in general give a picture of what I mean. For 'economic' reasons we have replaced manpower by all kinds of machines and now we wonder about increasing unemployment rates over much of the world.

People seem to know what they want. Why for instance do many prefer flats in old buildings? Certainly it is not only the bigger room heights or the nostalgic aspect. There are certain unnamed reasons which I think are unconscious feelings that something is wrong with modern buildings. Is it the density, the discomfort, the ugliness? Why do we say that it stinks or smells bad from fresh paint? Isn't there paint which has a good, pleasant odor? If we look closely at the contents of these products they often are poisonous and can affect the health of people. But the money involved in the building industry does not allow a 'quarantine' period before people move into a new flat. Usually there is only a check on the humidity before a new building is given over to the users. The fact that concrete for instance needs three to five years till it is dried out is generally overlooked.

And no one ever asks about the existence of toxic gases or radioactivity. For this I give the following example. In 1971 the State Laboratory for the control and inspection of food stuffs in Geneva was given a new functional building with all the necessary technical and sophisticated installations. As work started in the new building all the examinations resulted in toxic levels above tolerance. When control measures were made in the old place the toxic levels found were back to 'normal'. Finally it was found that the modern 'high-quality' materials in the new building were the cause of the high toxic levels measured in the food stuffs (and that after a short storing time only!). Toxic gases were leaking from paints, plastic materials, varnishes, flooring and furniture materials and so were poisoning room air and food stuffs to be examined. The scandal was soon forgotten and there was no change in the building regulations or the approvement rules for building materials.

If we build our houses of natural materials no harmful excessive positive oxygen-ions are formed. The inner surface of wood for instance is so large (unmatched by any artificial product) that it is able to absorb dust particles and neutralize bad odors. It also regulates the content of humidity in the room air. This is the reason why wood should be left as natural as possible and only be protected with a thin layer of linseed varnish or bees wax.

Brick walls are also excellent if they are built and plastered with a lime mortar in place of cement (cement mortar is a poor insulator and obsorbs moisture). When using bricks we must not forget that the initial energy to kiln dry them is very high. That is one reason why I consider the mud brick as one of the most ecological building materials. It is made from the soil on which the house is to be built. There is no energy wasted for the burning process or for transport. The material is fully recyclable. We also should not forget the healing effect of earth. Mud brick walls create a very healthy living climate. Mud fulfills most of the skin functions. It has an excellent heat storing capacity but is an insulator and has to be protected from humidity. For its production it has a very low input of energy but asks for more manpower. That could add on the other hand to its attraction in our present employment crisis!

THE 'BUILDING DISEASES'

It was in the fifties when the German doctor Hubert Palm found evidence that the cause for many of the ailments of modern man is found in buildings. That is why we simply talk about the following 'building diseases'.

CHEMICAL DISEASES

Synthetic materials for insulation, flooring, furnishing and finishing (including all kinds of plastics, chipboards, paints, etc.) usually set free all kinds of toxic gases which can cause headache, vomiting, loss of concentration, etc.

ELECTRICAL DISEASES

Electrical appliances and wires create electromagnetic fields with constant 50Hz frequencies within the house, often interfering with 16 2/3Hz from nearby railways or high voltage overhead lines and can cause nervousness or so called 'electrostress', etc. These artificial frequencies are in contrast to the 10Hz frequency of the earth and the human organism. Another phenomenom often observed in connection with synthetic materials is the electrostatic charge. The atmosphere is full of waves (radio, TV, radar, etc.) which interfere with the natural spheric radiation. We know that all biological processes work with the help of minute electrical impulses. Electrobiology is revealing new findings about the enormous influence of electricity to our body and health. One very effective solution to this problem is a central switch preventing automatically any current in the wires as long as there is no user. This is especially important during the night, when the sleeping organism regenerates and is more sensitive.

CAGE DISEASES

Concrete or steel buildings usually form a *Faraday Cage* within which life is dying out in the third generation. The loss of reproductivity which was found in laboratory tests with rats stems from the lack of the natural radiation from the earth and the atmosphere within the Faraday Cage. These natural waves which make life possible are sealed off or disturbed by reinforced concrete, steel pipes and the friction of running water within pipes, electromagnetic fields along electrical wires, etc.

DISEASE OF LOCATION

Underground currents or dislocations can be the cause of various chronic diseases like rheumatism, disturbed sleep and even cancer. In earlier times the check on a new building site by dowsing before construction was quite common. One can imagine that new construction sites are on disturbed or 'unhealthy' areas which were omitted by our ancestors for known reasons!

Since Building Biology is interdisciplinary or better co-disciplinary it deals not only with building materials and their influence on the human body but also with the environment in general, and with the climate of living which is determined by:

- installations and furnishings,
- noise, acoustics,
- light, lighting, colours,
- radiation, avoiding disturbed areas (dowsing),
- radioactivity,
- space, form and proportion,
- physiology of living and working,
- research methods,
- city planning, including its biological, ecological and sociological aspects.

Building Biology tries to cover all aspects of living in connection with shelter, a step forward in harmony and balance with nature!

REFERENCE

Rifkin, J. 1980
Entropy, A New World View
Viking, New York

AUROVILLE RESTORATION WORK

Auroville is an international community of some 500 men, women and children from more than 20 nations which was established in the early seventies under the inspiration of the spiritual teacher Sri Aurobindo. This collaboratively written report on land reclamation strategies was disseminated by Aurovillians in 1980.

Auroville covers some 11,000 acres, 2,000 of which are under direct use, scattered in small plots amid nineteen square miles of village, government, and temple lands on a low-lying plateau 100 miles south of Madras. This plateau, sloping down to bottom lands and stretches of beach along the Bay of Bengal, represents in many ways a typical terrestrial situation. Severely deforested, overgrazed and over-cropped, the land is subject to chronic wind and water erosion. Less than 10% of an annual rainfall of 4 feet is retained on unchecked land, and washout has robbed wide areas entirely of topsoil. Human and domestic animal populations already overstress local life-support capabilities. Increasing exploitation and decreasing restorative care inhibit the land's ability to sustain life at all.

Most years this area gets two monsoons. The southwest monsoon brings intermittent rain from May through August (rarely totalling more than 13 or 14 inches, just enough to allow the land to be prepared and crops started). Heavy rain – and occasionally cyclones – come with the northeast monsoon in October-December, when most of our rainfall comes down. January through May is predominately dry, with temperatures over 100°F in the latter part of the season. Gale force and heavier winds are common in the northeast monsoon season, and terrible hot, dry, dusty blows are frequent in May through July.

THE SUBTLEST OF CATASTROPHES

It is the subtlest of catastrophes, at least in these early and middle stages. If harvests are often inadequate, still the land does continue to produce crops; malnutrition may maim lives, but outright starvation is not here yet; under - and unemployment continue to increase, but still most small holding and landless families manage in some fashion to support themselves. It seems, in fact, as if life were going on as usual.

But life is not going on as usual here; it is slowly winding down. Traditional land use practices – monocropping during the wet season, free grazing by far too many herds during the dry – are part of the problem. Uncertain enterprises under the best of conditions, always under-financed, traditional farming practices leave little margin for regenerative action. In fact, neither traditional nor industrial agribusiness technology show any signs of competence in dealing with the kind of total environmental deterioration we're faced with here.

RESTORING AND PRESERVING

Our approach is to consider the environment comprehensively and apply restorative and preventive measures over as wide an area as possible. This means, for our area, water conservation work along with afforestation; water resources must be developed to sustain the trees through their first few

Large bund enclosing catchment area in the Greenbelt. (Photos, Auroville Greenwork)

hot seasons, and the soil stabilized so that planting work isn't simply washed away or silted under with the first heavy rains. We try to work down from the high points of the plateau with a network of earthwork microcatchments *(bunds)* and other small convservation structures (check dams, gully plugs, causeways, etc.) while setting in first a mixed plantation of drought-resistant shrubs, grasses, and trees, which must be protected by watchmen, fencing and thorn *baskets.*

Thorn baskets protect young trees.

These measures, employing the simplest technology with local hand tools and labour, work very well. Over the past few years, on land handled in this way, we've seen erosion come to a standstill and water tables rise.

It has to be emphasized, though, that none of these measures are of much use unless carried out as part of a comprehensive program: to build a dam, for

example, without prior erosion control work upstream, invites topsoil runoff and silting of the dam site.

Similarly, there's more to planting trees under these conditions than just sticking them in the ground. We lost thousands of trees in our first few plantings due to neglect of basic back-up and protection. Even the hardiest tree species have difficulty surviving the hot winds and dust and hordes of hungry goats of the dry season.

DEVELOPMENT BASED ON TREES.

This approach constitutes a holding operation, a way of arresting deterioration and preparing for rebuilding. Beyond lies the need to create a permanent life-support system, a self-perpetuating ecology/nomy. Thousands of people are trying to make a living on this plateau, and the land is simply too depleted to support them much longer through field agriculture. It seems from our experience that the only real possibility for continued growth and development here is through basing cultivation on and around trees.

Tree-based multilevel cultivation – integrated planting of a variety of ground, shrub, and tree species – opens up a fairly wide range of economic options. Food, fodder, fuel, shelter, clothing, cleansing agents, medicines, craft materials, almost any human need in this climate can be met by following the multi-level pattern of the forest.

This kind of land use is year-round and labour-intensive, affording much fuller human employment than the six months on/six months off cycle of monsoon agriculture. The trees not only shelter intercultivated crops from sun and wind and storm violence, they are themselves far more abundant and dependable producers (less susceptible to weather and pests, less dependent upon humans to survive) than are field crops. And the very presence of trees creates a sheltered and nurturing environment in which symbiotic, self-sustaining, and regenerative opportunities can arise.

Dry-farmed field crops (dependent on the monsoons for water) in this area include three kinds of millet, peanuts, various lentil-type legumes, a kind of coarse red rice and manioc. Irrigated crops include standard rice, vegetables, bananas, papayas and usually 'off-season' millets and legumes. The major tree crop here is cashews, which can be grown without irrigation and for which there is a ready export market. So ready is this market, in fact, that although cashew trees are everywhere, the high-protein nuts are rarely seen in the villages. Other tree crops include palmyas, coconuts, badam (Indian almond), mangoes, guavas, limes and a wide range of lesser known tropical fruits, berries and nuts.

A young casuarina forest.

From desert to forest in 7 years.

Using small, frequently hand or wind-powered pump sets, diligently bunded and forested and fenced, a handful of developing forest-farms now dot the plateau. Small plots of groundnut or millet alternate with stands of fruit trees and irrigated gardens of vegetables and flowers, bananas and papayas. Chickens and cows fit happily in, and algae, fish and methane gas production are being tried. We have barely begun, but we have done enough to see that a technologically simple but comprehensive approach to environmental regeneration is possible, that it works and is applicable to the situation in which we find ourselves.

GREENWORK IN THE VILLAGE CONTEXT

Until now, however, we've made very little headway in terms of enlisting local support. Cows and goats are still herded into our plantations, trees are still being cut, village lands are still deteriorating, fields are still being whitened with government-subsidized DDT, and confrontations are, if anything, getting more heated. There remains a widespread lack of local comprehension of the connection between cutting trees, erosion, declining soil fertility, and the life-endangering consequences all this implies. And even when comprehended, the hard facts of life in a marginal economy make regenerative options impossible for most villagers.

Where we see a protected young forest, a hope for the future, it's difficult for a goat-herd to see anything but encroached grazing land. And it's impossible to remonstrate with him, to suggest, for example, that he enter another more 'ecological' line of business, unless we have alternatives ready to offer. The small amount of work we've done so far does indicate that the range of employment opportunities inherent in a comprehensive program of environmental regeneration is very wide indeed. But the program is such that the bulk of the work is restorative and protective – not, in itself, economically productive. With restoration and production the environment *will* be brought to the point where it can generate economically productive activities, but to get to that point some form of continuing subsidy is required, for at least the first few years. With a few small grants and help from families and friends in the West, members of the Auroville manage to avoid the necessity of supporting themselves, and so are freed to work on making the land once again supportive. If a goat-herd doesn't have such subsidies, what is he to do but keep on herding goats?

The one really effective way to get the message across is through living example: once people can *see* the benefits of our way of working with the land in terms of increased crop yields, cash returns and the like, things sometimes begin to click. But the example alone is not enough; broad-based village cooperation requires not only further education and encouragement but a much wider distribution of land and capital than is now the case – in other words, radical social and political change.

Meanwhile, hundreds of people have found employment either working directly for Auroville or on government projects contracted to us. We offer free seedlings, help with planting, maintenance, and erosion control. And something is beginning to happen. An increasing number of local people are asking for and planting fruit trees and requesting our help in *bunding* their land. A few have come to live and work with us directly.

The extent of the work is massive; it's going to be a long haul. We'd like to be in touch with others involved in this kind of work.

Auroville Greenwork,
Unity Resources,
Auroville 605 101,
Tamil Nadu, India.

TREES AS ANIMAL FEED

By JASON ALEXANDRA

The recent drought has highlighted some of the inadequacies of grasses as the principal source of livestock fodder under Australia's conditions. While the grazing industry is still reeling from the effects of the drought and much of this country is suffering from the longer term consequences of over-clearing (salting of soils and waterways, soil erosion and loss of productivity) recent research, farmer and Department of Agriculture enthusiasm is paving the way to the widespread use of trees as reliable sources of animal feed for both regular and drought years.

Trees such as the Honey Locust *(Gleditsia triacanthos),* Tagasaste or Tree Lucerne *(Chamaecytisus palmensis),* not only provide feed for animals but are capable of multiple benefits to farms. These include stock, crop and water catchments protection, erosion prevention and restoration of fertility to nutrient depleted soils.

At present annual fodder production is dependent almost exclusively on annual and perennial grasses and legumes. The promise of highly productive, manageable pasture lands led to over zealous clearing of forests, and while the English pastoral system was appropriate under European conditions its application in Australia where droughts are not unprecedented disasters but regular occurrences has created many problems. Even in usual years the comparatively shallow rooted grasses are unable to grow throughout hot and extended summers.

In S.E. Australia both annual and perennial pastures reach peak production in spring, with autumn growth depending on early rains. The usual strategies designed to cope with this spring peak involve stock culling, the expensive practices of hay making – cutting, baling, carting and feeding out – or the use of supplementary grain feeds. It is possible to plant tree crops to supplement pasture and eventually make summer, winter and drought fodder available from a variety of trees.

This diversification into trees for fodder can stabilise farm production by buffering the variables of seasonal fluctuations and those of good and bad years. To stabilise annual fodder production tree fodder crops are used at times such as mid winter and mid summer, when herbaceous pasture is unable to supply sufficient feed.

Furthermore the high protein and concentrated nature of some tree crops enables the more efficient use of dry pasture or rank grasses by grazing animals.

S.T.C. Summer Tree Crop   A.P. Annual Pasture
A.T.C. Autumn Tree Crop   P.P. Permanent Pasture

Photos J. Alexandra

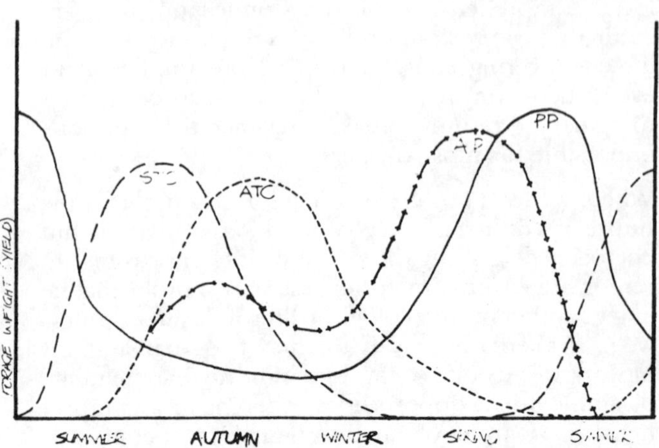

Schematic of even yield forage over year.

HONEY LOCUST

The Honey Locust is a tall fast-growing tree native to North America, surviving in all but really arid regions. It is well suited to much of Australia, tolerating both drought and frosts, with a minimum rainfall requirement of about 500mm (20") per annum, although growing better in higher rainfall districts. Its potential lies in the large pods which it can produce abundantly. These pods, between 300mm and 500mm (12-20") long, being rich in both sugar and protein are relished by livestock, including pigs, although they are also suitable for human consumption.

On good soils Honey Locust have been known to grow to a height of 30m (100ft).

They will grow on hillsides with little topsoil as they have deep, penetrating root systems. The canopy is open and graceful, allowing light to pass through to crops or pasture below.

J. Russel Smith, author of 'Tree Crops', reports that a three month old seedling only eight inches high had a taproot system 40 inches deep and widely spreading horizontal branches 6 to 18 inches long. These rooting characteristics make it superior for soil erosion control.

As the tree grows it produces large spikes on its trunk thus protecting itself from stock damage and only requiring protection while young. It cannot tolerate shade and when planted in an open, uncrowded situation forms many strong branches near the ground that are rarely damaged by wind, making Honey Locust ideally suited as a Windbreak tree. I would suggest planting shelter belt trees or crop trees at distances of about 5m (20'), culling inferior trees in about 20 years if necessary. In America between the years 1934 and 1943, 19,000 miles of shelter belts were planted on more than 33,000 farms as part of the "Prairies Shelter Belt Plan". Of all the species planted, the survival rate of the Honey Locust was the highest at 79.3%.

The pods ripen in Autumn and drop to the ground where they can either be collected (raked), (Hay rake) or directly consumed by stock. The Alabama Agriculture Experiment Station reported in 1942:

"5 year old trees of Milwood Variety produced an average of 58.3 pounds per tree (dry weight) which at 48 trees to the acre would be the equivalent of 2,798 pounds of pods...a feed equivalent of about 90 bushels of oats. The 1946 report states: ...some of the 8 year old trees produced over 250 pounds per tree. With 35 trees to the acre this would be 8,750 pounds of concentrate or the equivalent of 275 bushels of oats per acre."

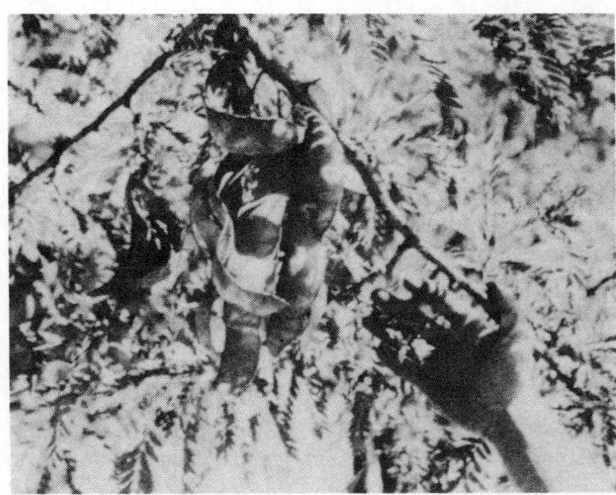

The open canopy and deep taproot on Honey Locust ensure that it does not inhibit pasture growth. Honey Locust pods rich in sugar and protein, fall in Autumn, dramatically increasing available fodder.

Seedlings cannot be expected to yield as consistently as selected cultivars but there are many reports of seedlings bearing generous crops with great regularity. Obviously yields increase as the tree matures.

During the same series of experiments above, farm animals were fed solely on Honey Locust pods and the results were excellent. An increase in butterfat and milk flow was observed as a result of the pods being added to the diet of dairy cows.

Mr. Trevor Leonard of N.Z. planted one acre of Honey Locust seedling 22 years ago. As a result he claims to have doubled production from that area with both grass and pods available to his cattle. He has won prizes for his stud bulls who are said to race each other as they hear a pod rattle down through the trees. He has recorded the highest documented yield with 660 lbs. of pods from a 20 year old single tree. His trees are expected to continue increasing in yield for the next 50 to 100 years with no additional costs.

DESCRIPTION OF HABIT

Deciduous, Fast-growing, shade intolerant, long lived to 120 years or more. Open crown with spreading branches casts light to moderate shade. Late to leaf out. Strong taproot and profusely branched root system. Pods narrow flattened 6-18" long; many seeded, dark brown, twisting when mature in autumn, hang on trees until early winter. Pods 20-40% by weight total sugars, 10-11% by weight protein (improved cultivars).

CLIMATE

Average annual rainfall in native range 20-70".

Temperatures range from 30 degrees to 100 degrees F. Growing season 140-340 days. Drought resistant once established. Wind and frost tolerant. Best in areas with sharply defined seasons.

SOILS AND TOPOGRAPHY

Commonly found on moist alluvial flood plains and limestone soils. Tolerant of wide range of soils. Tolerates acid, alkaline or saline soils. Upper altitudinal limit for best growth 5,000 ft. The only two major insect pests of the honeylocust are not found in Australia.

YIELD

Seedlings bear fruit pods from 6-10 years of age; improved cultivars at 5 years. At Alabama Exp. Stn., "Millwood" yielding 58.3-65.6 lbs/tree at 5 years of age and 250lbs/tree at 8 years. Five-year old "Calhoun" trees bore 26.4 lbs. Highest recorded yields 600 lbs/tree from 20 year old trees in New Zealand. Yield per acre (at 40-48 trees/acre) 1-12 tons, extrapolated from measurements of individual trees.

USES

Pods valuable livestock fodder; nutritional value by weight comparable to oats. Also important to deer, small mammals and some birds. Useful for erosion control, hedgerows and shelterbelts. Flowers produce pollen and nectar for honeybees. Thornless cultivars of G.t. inermis widely used as ornamentals.

Reddish brown heartwood tough and hard but sometimes brittle; decay resistant. Useful for fenceposts, decorative hardwood and firewood.

TAGASASTE

"If you were promised a permanent pasture which grew four metres tall on poor soil; was capable of producing more protein rich green feed than any known dryland crop – and which remained productive for thirty years or more, would you believe it? Read on!"

Tagasaste appears set to have a big impact on livestock production and farm environment protection in the next few years. Dr. Lawrence Snook a former Animal Scientist with the Department of Agriculture in W.A. has been able to triple the carrying capacity of his pastures at his Margaret River farm through the planting of 800 Tagasaste trees to the hectare. He is now able to stock 300 sheep per hectare while the district average is 10 per hectare. After years of encouragement to do so Dr. Snook now believes that many more farmers are ready to follow his example and plant Tagasaste. Field trials by Dr. Snook have found that a hectare can yield in excess of 11,000 kg of high protein dry matter annually. Leaf and edible stem have a crude protein content of

Tree Lucerne (Tagasaste) can tolerate heavy grazing and regenerate within months.

between 18% and 25% and digestibility of between 60% and 80%. This compares favourably to lucerne to top quality rye/grass clover hay. A hectare of Tagasaste trees is capable of yielding the equivalent of more than 440 fifty lb. bales of top grade feed. This is a higher yield than any other known non-irrigated fodder crop including lucerne.

Tagasaste thrives throughout S.E. Australia on a wide range of soils and climates. Many older plants can be seen growing on roadsides where they can get established away from grazing animals. It is a small to medium sized tree (4 to 8m high) with a dusty blue green foliage and masses of white flowers in July, August and September. It is a native of the Canary Islands where it grows on hot dry volcanic slopes. For centuries it has been used by the local farmers who take two cuts a year to feed their animals. It was first introduced to Australia late last century to be tried as a fodder plant. An article in the "Australian Pastoralist Review" October 15, 1900 praised its rapid growth and suitability as a hedge-row tree. It stated: "Let the new fence be put up to either side of the old one, with a gap sufficient for planting a hedge between them. In time the trees will grow to twelve or fifteen foot in height. If conditions are in any way suited to them..." it was these early plantings which account for the large numbers of wild trees which are growing throughout Southern Australia.

NOTES ON ESTABLISHMENT

-Tagasaste requires soils which do not waterlog. Other than this it will tolerate virtually any soil. It has a deep penetrating root system and is very drought tolerant. It is therefore well suited to light soils or hill country where it thrives on these otherwise low productivity sites.

-Minimum rainfall requirement is 300 mm according to Snook and it thrives in high rainfall areas such as the hill country of Vic., N.S.W., Tasmania and the Atherton Tablelands.

-Young trees must be completely protected from stock for at least two years to allow them to get sufficiently established before browsing. Under poorer conditions three years growth before browsing may be required.

-Growth rates often exceed 1.5m in the first year and frequently 2m or more in the second. Then growth slows down, however pruning and browsing stimulates rapid re-growth.

-Growth tips should be removed from young trees to encourage more branching and therefore greater fodder production.

MANAGEMENT STRATEGIES AND TECHNIQUES

When sufficiently established Tagasaste can be browsed by sheep all year round once the right stocking rates are determined. Alternatively larger numbers of stock can be moved into plantations when feed and shelter are required, i.e. at lambing, mid and late summer, during droughts or in winter and cold weather. Sheep will only eat as high as they can reach and therefore upper branches must be lopped manually or mechanically. Branches can be fed out to stock in adjacent paddocks or simply left for animals to feed from the ground. Tractor mounted hedging machinery is available which is ideal for this work, however Dr. Snook does all his lopping with long handled clippers. He considers it easy and pleasant work compared to the time, effort and money required to cut, store and feed out hay.

Lopping when required allows a block of Tagasaste to fulfill the same function as a hay shed or silage pit without the bailing and storage requirements.

Cattle and other large grazing animals must be managed more carefully than sheep. Farmers can use strip or intermittent browsing to prevent long term damage to the trees. It is easy to tell if stock are overgrazing a stand by checking to see if the bark is intact.

Another area of large potential is the use of Tagasaste as a supplementary summer green feed for the dairy herds when keeping milk quotas up is all important.

To maintain high yields Dr Snook uses annual dressing of superphosphate or super potash 5:1 at approx. 200 kg per hectare. Contrary to this, research from N.Z. showed that Tagasaste grown on dredge tailings from the Taramakau River was more productive without fertilizer.

FODDER PLANTATION LAYOUTS

Various densities of plantations are dependent on proposed management techniques.

Low density agroforestry. With this system wide spaced rows of Tagasaste are intersown with permanent pasture or crops. Rows approximately 10 metres between trees require approx. 400 trees per hectare. Such a system allows for strip grazing, mechanical harvesting and pasture or crop growth in the sheltered inter-row space.

Medium Density – 400-800 trees per hectare or a 5m x 5m to 5m x 2.5m grid. This still permits mechanical hedging, vehicle access between rows, and some grass growth. Higher densities give greater yields. Plantation blocks make ideal lambing paddocks with their abundance of feed and shelter.

FODDER WINDBREAKS

The rapid growth of Tagasaste makes it ideal for quick, easy to establish windbreaks. These can be used to protect gardens, pasture, stock, orchard and horticultural crops. These windbreaks can also be used as fodder reserves. A double windbreak row of Tagasaste planted 2.5m apart, one kilometre long would contain 800 trees and therefore be equivalent to a one hectare plantation. Gateways into such windbreaks could be used to let stock into browse or provide vehicle access for slash harvesting.

In Victoria's Western District tree lucerne hedges protected from overgrazing by netting provide a protein rich supplement to poor quality pastures in Autumn.

OTHER USES FOR TAGASASTE
Nurse Shelter

Tagasaste with its rapid growth, ease of establishment, tolerance of wide ranging conditions and ability to improve impoverished soils, can be a major pioneer species for the subsequent development of more diverse tree cropping on farms.

As many of the important crop trees are longer lived forest trees their relationship to pioneer species is of great importance. Pioneer species create the soil conditions and provide the shade and shelter which forest trees need in order to flourish.

Take as an example a farmer with a long-term policy of diversification into nut and timber production to augment grazing. Starting with a program of Tagasaste plantings on all steep and stony land 10% of the total is planted every year for ten years.

By year three the first Tagasaste fodder block is yielding well. By year eight, five blocks are old enough for browsing. The farmer then begins inter-planting nut and timber trees in the first tree lucerne block. These young trees are benefited by eight years of soil improvement carried out by the deep-rooting nitrogen fixing legume and thrive in the sheltered, mulched conditions.

Until the forest trees are sufficiently large to cope with stock the farmer is still able to use the cover crop of tree lucerne as a slash and carry fodder reserve.

Promising results have been achieved in New Zealand using Tagasastes at very close spacings one to two metres from walnuts and Chile nuts. Those sheltered with Tagasastes recorded 15% and 20% increases in growth over those growing without shelter. The use of nurse shelter is an important way of gaining straight boles in speciality timbers such as Blackwood and Black Walnut.

Bee Fodder and Honey Production
The early and extended flowering of Tagasaste and its high nectar and pollen content make it an important winter-spring feed for bees.

Pollination
Attracting bees to an orchard, crop or garden can increase production by greater pollination.

Fire Resistance and Protection
Tagasaste has low flammability. It stays green all summer and does not burn readily. It makes good fire breaks.

Windbreaks
As already stated, it makes a rapid bushy windbreak in a couple of years.

Most larger farming properties have areas such as the steep, stony or eroded slopes, cold or windswept ridges or waterlogged spots which can be planted to trees without taking prime land out of production. Tree crops can further increase productivity on these marginal sites. Various species grow and produce fodder under different conditions, e.g. Willows and poplars thrive on wet or swampy ground. Carobs, kurrajongs and tree lucerne grow on dry slopes.

In wind ravaged New Zealand, Peter Smail lost 250 lambs in one night to exposure. Since 1953 he has doubled his stocking rates by planting thousands of trees for windbreaks lambing havens and woodlots. The drop in wind velocity protects lambing ewes and off shears sheep and cuts moisture loss from evaporation and transpiration in pasture, promoting growth. A flock of range-reared geese and turkeys shelter in the woodlots by night and feed on insects and seeds in adjacent paddocks during the day.

Trees give effective shelter across a paddock to a distance about 10 times the height of the trees. Pasture and crops in this protected zone show better growth rates. Deep ripping alongside tree cuts roots that compete with crops for moisture.

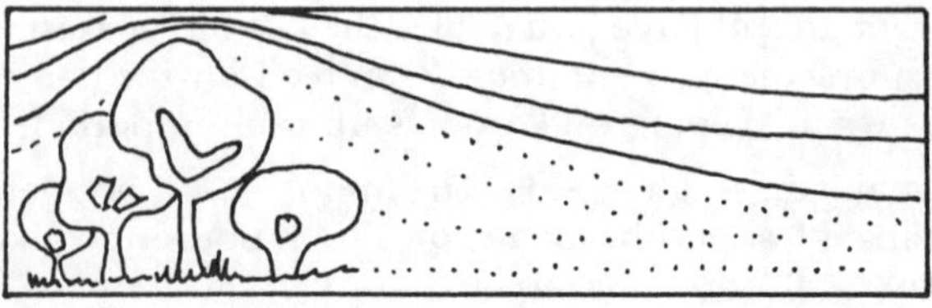

Tree clumps give more effective shade and shelter than single trees spread throughout a paddock and natural regeneration is assured.

Source: 'Tree Lucerne': an historical perspective and discussion of potential uses in New Zealand.' P.J. Davies, Crop Research Division, D.S.I.R., Private Bag, Christchurch, New Zealand.

FIGURE 1: USES OF TAGASASTE (TREE LUCERNE) — FOR NEW ZEALAND

Uses	Pastoral	Forestry & Agro Forestry	Horticulture	Cropping	Conservation	Miscellaneous (industrial)
NURSE SHELTER	**	**	**	**	**	Nil
WINDBREAK	**	**	**	**	*	*
SHADE	**	**	*	*	**	*
FERTILITY N	**	**	**	**	**	*
HONEY	**	**	**	*	**	*
BEE FODDER	**	**	**	**	**	*
POLLINATION	**	**	**	**	*	*
ANIMAL FORAGE	**	*	Nil	*	*	Nil
SEED PRODUCTION	*	*	*	**	*	**
PELLETS (Stock Feed)	**	Nil	Nil	**	Nil	**
CONCENTRATE	*	Nil	Nil	*	Nil	**
BIOMASS	*	**	*	*	Nil	**
FUELWOOD	**	**	**	**	**	**
FIRE PROTECTION	**	**	*	*	**	Nil
WEED SUPPRESSION	**	**	*	**	**	Nil
WILDLIFE	**	**	*	*	**	*

NIL   * MINIMAL   ** SIGNIFICANT

PROTEIN

The crude protein (CP) content of Tagasaste in Western Australia and New Zealand compares favourably with the better quality pasture forages of New Zealand. However, this is dependent on the proportion of stem. A ruminant animal requires 70 g CP/kg DM for maintenance so protein is not likely to be a limiting nutrient.

TABLE 3: Crude protein content of some browse plants and New Zealand forages.

Source: D.P. Poppi, Department of Animal Science, Lincoln College, Canterbury, New Zealand.

	Plant Part	CP (g/kg DM)	Country	Source
Browse plants				
Tagasaste	leaf and edible stem	180-250	Western Australia	Snook (1961)
	plant tips	235	N.Z.	Davies (unpublished data)
	leaf and stem up to approximately 8 mm in diameter	178	N.Z.	Poppi and Davies (unpublished data)
	leaf	226	N.Z.	
	stem (up to approx. 8 mm in diameter)	81	N.Z.	
Leucaena (*Leucaena leucoephala*)	whole plant	125-243	—	Skerman (1977)
Pigeon Pea (*Cajanus cajan*)	whole plant	115-214	—	Skerman (1977)
Mulga (*Acacia anerura* F. Meull. ex Benth)	leaves	117-132	—	Skerman (1977)
Pasture and Hay				
Ryegrass/White Clover				Ulyatt *et al* (1980)
(spring)		220	N.Z.	
Lucerne (immature)		250	N.Z.	
Lucerne hay (pre bloom)		200	N.Z.	
(weathered)		120	N.Z.	
Meadow hay (young leafy)		120	N.Z.	
(mature)		100	N.Z.	
Barley straw		40	N.Z.	

Some 3000 range-reared pigs are housed in this East German forest during the summer months, reducing sunburn worries and producing leaner pigs as a result.

At Cottonwood Farms in the Hunter Valley of New South Wales, Australia, cattle production has increased 20 percent under wide-spaced, high pruned poplars grown for the quality furniture timber market. Prunings provide nutritious leaf fodder higher in protein and minerals than lucerne. In the establishment phase pumpkins and passionfruits were grown between the young trees.

Direct seeding of trees for broadacre plantings is being pioneered in Victoria's Western District. This machine incorporates an opening ripping tyne followed by two angled cutting discs. The seed is coated with nutrient powder and mixed with a pollard/bran mixture. Directly sown trees have healthier root systems and greater vigour at much reduced costs.

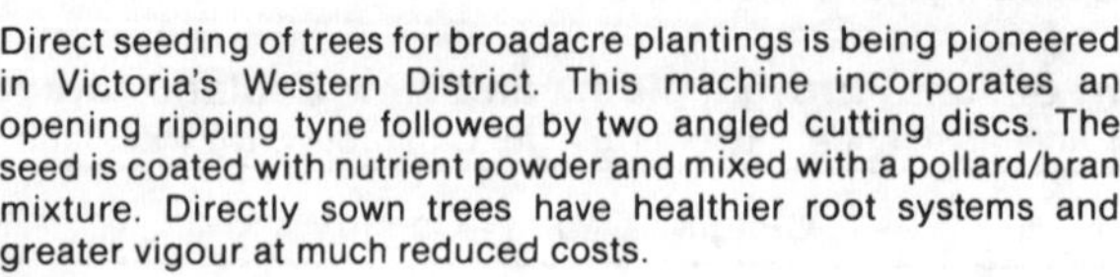

In Southern Tasmania a crop of 325 bales of hay per hectare was cropped from this agroforestry paddock; about the same as before it was planted with pines five years ago. After shearing and during lambing the paddocks provide excellent animal shelter, reducing mortality rates.

Kurrajongs dotted over wheat fallow near Parkes, N.S.W. are used for shade and fodder.

In Australia 2.2 million square miles of arid and semi-arid land supports nearly 50 million sheep and 4.5 million beef cattle.

Massive livestock losses are sustained because of climatic stress on unsheltered lambs, calves and off-shears sheep.

Extremes of heat and cold have dramatic effect on animal weight gains and reproductive rates. Paddocks, yards and watering places require shade plantings suited to climatic conditions.

The retention of native vegetation on steep erosion prone hills complemented by productive introduced tree species in on-farm plantings create a stable ecological framework for long-term farming.

Fencing of solitary trees allows natural regeneration to take place.

In many areas of Australia where suitable native trees exist, these have been used as drought fodder. However, the conscious integration of fodder trees into farm planning has rarely been instituted. The abilities of windbreaks and shelterbelts to improve the productivity of farming properties is well known. To go one step further and actually use trees and perennial shrubs as a prime source of fodder has rarely been attempted. The attempts that have been made have shown that there is enormous potential for harvesting fodder from several tree and shrub species. The species yielding fodder that have proven successful in part of S.E. Australia include:

MULGA AND TREE WILGA
(Acacia aneura and Geijera spp.)

Both of these trees grow naturally in inland areas and hence are suitable for dry regions. Both can be slashed to provide fodder and the trees recover well from pruning if sufficient branches are retained above the reach of grazing animals.

KURRAJONG
(Brachychiton populneus)

A widely distributed native throughout sub-coastal areas of south of the Tropic of Capricorn. Deep rooted and capable of surviving in dry areas, it has been widely planted as a street tree in inland Victorian towns. The foliage is slash-fed as emergency fodder and the tree responds well to pruning.

WILLOWS AND POPLARS
(Salix spp. and Populus spp.)

The Weeping-Willow and several Poplar species have foliage suitable for fodder. Their tremendous growth rates may make them an extremely good way of maximising fodder production in wetter areas. A vigorous root system makes both trees valuable also for erosion control.

MULBERRIES
(Morus spp.)

The leaves of the Mulberry are very palatable and nutritious. It requires well-drained soils and if annually pruned can be a very useful fodder source.

OAKS AND CHESTNUTS
(Quercus spp. and Castenea sativa)

These trees yield nuts useful for pig fattening, deer fodder (especially acorns, which they find particularly palatable) and also for human food. Although species such as the evergreen and holm oaks grow in relatively dry areas, most species prefer higher rainfall areas and are drought tolerant once established.

CAROB
(Ceretonia siliqua)

A tree widely grown around the Mediterranean for its nutritious and sweet pods. It is a drought hardy species suited to a semi-arid area. Pods are borne on the female trees only and apart from the fodder it is a useful shade and shelter tree.

REFERENCE

Smith J.R.
Tree Crops: a Permanent Agriculture
Devin - Adair. Old Greenwich 1977. 4th printing.

A LEUCAENA BASED CATTLE FORAGE SYSTEM

By LEIGH DAVISON

I first became interested in Leucaena leucocephalla some three years ago when I read an article describing its use in the rehabilitation of eroded and degraded land in the Phillipines. In this case the K8 (Hawaiian Giant) variety was being used. Some months later a friend showed me an article in a Queensland DPI Journal which described the use of bush varieties of leucaena as the basis for a cattle forage system in the Rockhampton area (23.22 S, 150.32 E).

In this system, leucaena was planted in parallel rows about three metres apart, and allowed to grow for several years with little or no grazing. By this time the trees were large enough to withstand serious damage from grazing cattle, and were mature enough to have produced several crops of seed. In the situation described, this seed had been allowed to drop and germinate in the spaces between the rows of mother trees. It was the carpet of leucaena seedlings that provided the main forage. The mother trees provided some forage from lower branches, but their main role was to provide seed for the replenishment of the seedling carpet.

With the intention of trying to emulate this system, I purchased in August, 1980 1 kg of *leucaena peru* seed (*Cunningham* was not then available in commercial quantities). That same month I had a bulldozer clear and contour-rip an acre of weedy hillside on our property at The Channon in the Lismore area of northern N.S.W. (28.48 S, 153.17 E). The soil type on the plot is red to chocolate-red basalt. The aspect is north-easterly. The weed cover varied from relatively luxuriant lantana on the gentler slope, to crofton weed and poor quality lantana in the steeper areas.

As there was no rain during September and October, 1980 I was not able so sow the seed until the first week of November. I scarified the seed with boiling water and innoculated and lime-pelletized it. Because the dozer had left the area clean, and the soil friable, I was able to sow simply by dropping the seed into the furrows left by it.

The author watches 'Fairy Faith' work a leucaena bush over. (Photo, Leigh Davison)

I applied a couple of bags of dolomite, believing, at that time, that leucaena required alkaline conditions. I have since learned that any pH over 4.5 is acceptable. A bag of rock phosphate was also applied. As a cover crop I sowed cowpeas very thinly in the intervening furrows. This was a mistake, as the vigorously running cowpea vines were soon growing all over the leucaena seedlings, and I was kept busy all summer with the brushhook.

The leucaena struck well, and was helped by regular rainfall throughout the summer-autumn period. The strike was poorest on the steeper slopes where growth was also slower.

It was interesting to note that there was little or no strike in the vicinity of a flooded gum which I had left standing. Those plants that did emerge near the gum tree have fared badly, indicating the existence of a leaf or root exudate that inhibits germination and growth of the leucaena.

After a year the better specimens had reached about 3 meters in height, with stem thicknesses up to 1.5 cm, and were flowering. The trees set a lot of seed, and I was hoping for a good secondary strike of volunteer seedlings in the spaces between the rows of mother trees. The wet season of 1982 was even better than that of the previous year, so my hopes were high.

85

However there was little or no germination. I had noticed that many of the seed pods contained a white fungal growth and this must have rendered most of the seed non-viable.

The trees made good growth throughout 1982 with the customary pause during winter. In January 1983 I decided to slash the trees to about 1 meter in height so that the resulting regrowth would be within reach of browsing cattle. I performed this operation with a brushhook, and was followed down the rows by several eager cows. It was interesting to note that they preferred the succulent green pods to the leaves, which they also relished, stripping the limbs bare.

During this slashing operation, I noticed that the mature seed was generally free of the fungal blight so prevalent the previous year. This was no doubt, due to the remarkably dry conditions in January 1983. It was, in fact, so dry that on that red soil slope the leucaena had dropped virtually all their leaves by the time the first rains came in mid-February.

A couple of weeks after this rain I returned to see how much leaf growth had occurred on the stumps of the slashed trees. To my delight, not only was there considerable growth in evidence (in contrast to the trees which I had not yet slashed), but also the long awaited carpet of secondary volunteers had begun to emerge in the areas where slashing had spread and opened the seed pods.

During the 2.5 years of this experiment I have had to spend time chipping and brushing weeds like wild passionfruit, wild tobacco and mistflower. This has had the effect of encouraging the reappearance of several grass species including Kikuyu which is, no doubt, benefiting from the excess nitrogen produced by the leucaena. The original ripping operation caused a solid germination of some dormant siratro (*Nacroptilium atropurpureum*) seeds. This vigorous running legume tends to cover the trees and pull them over. I have controlled this siratro by periodic grazing, as the cows prefer it to the leucaena.

From this experience it would appear that the rate of germination of the secondary crop of leucaena, so necessary for the realisation of the forage system's full potential, depends heavily on the amount of rainfall occuring in January when the seed is hardening up. January 1982 produced 350mm while January 1983, the year of high strike, saw only 50mm of rain fall.

I have celebrated the breakthrough by treating the area to a bag of blood and bone, to maximize pre-winter growth of the understorey carpet.

I am planning to put in a plot of several acres in spring of this year. It will be on flatter land, so that weed control will be possible using my walk-behind type sickle bar mower, instead of hand tools. The

Leucaena laiden with pods.

initial fertilizing will be with blood and bone (N & P) to encourage more rapid growth. I will probably sow grasses and pigeon peas inter-row as protective cover until the secondary germination occurs, at which time a light cultivation may be advantageous.

ADDITIONAL NOTES FROM LEIGH (January 1985)

1. Due to heavy weed competition between the rows of mother trees I decided to plant grass rather than rely on the secondary seedling strike. This has worked well and a good sward of pasture grasses (broadleaf paspalum, Rhodes grass and setaria) is now well established between the rows of leucaena trees. The trees are all pruned back to about one metre above the ground and the cows are let in every month or two, before the coppicing shoots grow out of their reach. This eliminates the need for continued pruning.

2. It is interesting to note that in orchards where leucaena has been planted extensively as a green manure source the 'seedling carpet' effect is working well and is providing a leguminous ground cover that competes well with weeds (in the absence of selective grazing). In the orchards the leucaena is pruned twice per year in summer and the leaves piled around fruit trees. Depending on the state of the seedling carpet, some trees can be left to seed in order to replenish the ground cover.

3. I recently had a soil test done. Our pH turns out to be about 4.2, very acid. Work in Brazil on very acid red soils has shown that leucaena responds to gypsum much better than to dolomite or lime.

I will be testing the response of my leucaena to gypsum. I will be somewhat cautious in view of the

Broadleaf paspalum ground cover in the foreground. 'Poppy' prefers succulent leucaena tips. (Photo, Leigh Davison)

opinion expressed by Allan Walker (1983). "Although gypsum contains calcium, it also contains sulphate which inhibits earthworms and microbes and therefore is not a good source of calcium."

I would be very interested to get feedback from readers on this question. What is the best way to get calcium down to the deep rooted leucaena plants?

REFERENCES

"IPIL IPIL" Spring/Summer of 1979
Permaculture Quarterly, page 30

Queensland DPI Journal, May/June 1980, Vol 106 No. 3

Walker, Allan
"The Miracle of the Soil"
Permaculture Nambour Newsletter, Spring 1983, page 14.

FURTHER READING

Leucaena, Promising Forage and Tree Crop for the Tropics
National Academy of Science
Washington D.C., 1977

REALISING THE MULTIPLE VALUE OF TREES

"WE OFTEN HAVE TO MAKE A CHOICE. WE CANNOT USUALLY EXPECT TO REALISE THE FULL POTENTIAL OF MORE THAN ONE VALUE"

By CHRIS VAN KRAAYENOORD

Chris van Kraayenoord is scientist in charge of the National Plant Materials Centre at Aokautere, New Zealand. The following report on the multiple values of trees is from Chris's introductory address to the 9th Annual Conference of the New Zealand Treecrops Association, April 1983.

I have always found some difficulty in coming to terms with this concept of multiple values and multiple purpose. You know what it means – a tree yielding not only a primary product but also having a supplementary value. If the most important value of a tree is its potential for wood production, what can be the second or third value? We can look for this extra value in a second product produced by the tree, whether it is nuts, fruit or fodder. However, the second product should ideally be available without detriment to the first product, yet this is not often possible. Usually we have to sacrifice one for the other. It may be possible to combine, for instance, wood production with a crop of nuts, such as walnut or chestnut, but to maximise the crop we have to sacrifice the wood value. Or if we want to obtain the maximum fodder crop, say from honey locust, we have to sacrifice the wood potential because we want the tree to develop many spreading branches with many, many pods. The point is that we have to be very clear in our minds when planting a multiple value tree. What is the product we are aiming at? What is the value we can most profitably realise? And we often have to make a choice.

If we plant trees for soil conservation, we should realise that we often plant on badly eroded, infertile sites. Such trees may grow under stress with minimal attention, and whilst it may be possible to get timber from them, it will never be very valuable timber, being only firewood, farm lumber or possibly pulpwood. To produce clear wood we need to grow quality trees on the better classes of land with regular pruning and management, to yield a valuable wood crop. Combining soil conservation and wood production may result in less valuable timber and increased erosion during harvesting and replanting.

To combine wood production and shelter in a single row shelterbelt raises other problems – it is possible but difficult to combine the two, and requires considerable management skill. However, systems have been devised, for *Pinus radiata* for example, of pruning every second tree, which prevents the belt

C.W.S. van Kraayenoord (Photo ERTI, Hungary)

becoming too dense and enables eventual extraction of the pruned tree for quality timber.

People may imagine that nut crops could be combined with shelter by interspersing nut bearing trees in shelterbelts or growing a row of hazelnut trees. The result may be shelter, but probably very few nuts, and in the case of fully exposed trees, probably neither. So that where a tree has multiple values we must define our priorities. We cannot usually expect to realise the full potential of more than one value.

Where the prime value of a tree is a crop, the tree has to be managed and protected to maximise the crop. Protection from wind damage and animal damage are essential. Many a treecrop has failed because insufficient shelter was provided. Plant crops before the shelter is established, or try to grow crop and shelter at the same time.

All crop trees need shelter – *Eucalyptus spp, Juglans nigra* and *Acacia melanoxylon* the special purpose timbers, nut and fruit trees – they all need shelter. If we plant for bee forage we also need shelter. Stock

88

fodder trees such as honey locust, also need shelter. So we have to pay much greater attention to selection, establishment, and management of our shelter trees.

Most tree crop plantings are in grasslands, often in the presence of sheep or cattle. Many schemes have been started with great enthusiasm, but the difficulty in giving adequate protection to the trees from grazing or from damage by stock, at least for several years, has led to many failures and disappointments. Stock protection is essential and protectors must be properly maintained. In addition, grass growing around trees competes for moisture and nutrients and can severely check growth, especially for the first few years after planting out. For optimal establishment and juvenile growth rates, it is essential to suppress grass competition.

I WILL NOW CONSIDER A FEW EXAMPLES OF USING TREES WITH MULTIPLE VALUES.

The first example is the growing of trees to produce timber, but the multiple use of the plantation for stock grazing as well. At Tikitere, near Rotorua, the New Zealand Forest Service and the Ministry of Agriculture and Fisheries have a combined trial researching the feasibility of growing Pinus radiata for timber at spacing far enough apart to permit sheep and cattle to graze around the trees. At Esk Forest there is a similar system under trial at generally wider spacings. A lot of money and good skill will be required to grow good timber trees under this regime, but if successful, it certainly could have useful applications. The same could be done with other tree species. Eucalypts and pasture as a possibility; also Acacia melanoxylon in sheltered valleys.

Wide spaced poplars planted for erosion control or hillside stabilisation may also yield timber. Poplars, are, of course, generally readily established in pasture as poles protected by plastic sleeves in the presence of grazing sheep. Poplars would have to be pruned up to produce clear wood. A farm well planted with wide spaced trees for soil conservation has considerable aesthetic appeal, such as is seen at Wharekiri, the former soil conservation demonstration farm near Gisborne, where many of our poplars and willows were tested.

In a walnut plantation in Victoria, Australia, cattle are allowed to graze the grass between the trees. The wide spacing of the trees permits multiple purpose use. But you still have to be careful and watch those cows!

Pine trees can have multiple values other than timber. Occasionally in France and the tropics they are tapped for gums and resins. In Korea people collect the cones of *Pinus koraiensis* from which edible kernels can be extracted for human consumption. Also in Korea, chestnuts are frequently planted as a food source. Often they are planted on eroded sites and so also serve a soil conservation function. Even these people are very careful not to plant chestnuts higher up the slopes, but only on the lower slopes where there is shelter and the soil is more fertile.

Wide spaced planted honey locust can combine the production of a tree fodder crop of pods, with grazing for sheep or cattle, but the trees will not be suitable for timber, although possibly they will provide some firewood.

Plane trees are similar. Wide spread planted trees help stabilise hillslopes, provide shade, and are a source of emergency fodder. You will not get timber from them, and to give stock access to the leaves you still have to lop the branches, so there is a lot of physical work involved.

Willows are also a source of fodder in times of drought, but to lop the trees is, again, a very strenuous job. It is much easier to grow the willows in a densely planted fodder plantation where the plants are browsed or cut back to the base each year. The sheep walk through the plantation browsing at leisure – the sheep do all the work and branches do not have to be lopped down.

We conduct similar trials with fodder shrubs. In the course of evaluating shrubs for soil conservation we need to determine those that are palatable to sheep and those that are not. *Atriplex halimus,* a fodder shrub for semi-arid conditions, has been used for several years in the South Island in browse trials. *Atriplex canescens* is another rangeland saltbush suitable for fodder.

I feel there are great possibilities of using *Robinia pseudoacacia* for the multiple values of timber, shade and honey production. Some people regard the thorns of black locust as undesirable – I agree, but I can tolerate the thorns in return for strong durable timber. Last year in Hungary, I was very impressed with the eight or so superior *Robinia pseudoacacia* clones selected by the Hungarian Forest Research Institute, and arranged for the importation of root cuttings by our Plant Materials Centre. These cuttings have been received and successfully struck – and we are now able to multiply up those Hungarian selections for evaluation in New Zealand. The selections include clones with a longer flowering period and nectar of a higher sugar content. There are about 30 further selections being developed in Hungary.

Robinia viscosa is a shrubby locust with very good honey producing flowers, but there are very few trees of it growing in New Zealand.

In conclusion, there is one point I would like to make referring particularly to tree fodder crops. We have made much progress in the work of Doug Davies with

The use of *Leucaena leucocephala* to reclaim highly aluminous soil, Weipa, northern Australia. The site is on a former bauxite mine, and the leucaena is being tested as a nurse crop for African mahogany (*Khaya* species), which is planted between the leucaena trees. (COMALCO, Weipa, North Queensland, Australia)

Black locust (*Robinia pseudoacacia*) growing vigorously on extremely steep, infertile mine spoil, Raleigh County, West Virginia, USA. After 4 growing seasons the trees have reached 4 m in height. (Soil Conservation Service, U.S. Department of Agriculture)

the tree lucerne, and that of Joan Radcliffe who is also working on tree-lucerne and on the tree medick. The Plant Materials Centre is working on the willows, and is assisting other research stations who are also working on willows for fodder. I think it is now time that we had trials on a larger scale. I would like to see a trial unit established of some 10 to 15 ha where not only willows, but also tree lucerne, tree medick and honey locust could be tested on a proper management basis so that we can find out the best utilisation of the crop and remaining pasture. We also need to know the effect of the tree fodder diet on the stock, and plant longevity under intensive management systems.

The National Plant Materials Centre is the main Centre for soil conservation research in New Zealand.

The objective of the Plant Materials Centre is to evaluate, develop and release improved plant materials for land stabilisation, erosion control, river control and shelter. It also carries out research into multiplication, establishment and management techniques, and diseases of soil conservation plants.

At Aokautere the centre maintains a 32 ha nursery and trial area.

40.20 S 175.39 E

Soil type
Manawatu silt loam
Manawatu fine sandy loam

Climate
Mean annual rainfall: 995 mm (at DSIR, 4.5 km distant)
Mean annual min. temperature: −2.7 degrees C
Mean annual sunshine hours: 1794 hours
Mean annual wind speed: 10.1 km/hr

BLOCK 1
Grazing trial to evaluate several shrub species, willow and poplar clones to provide animal fodder during periods of drought or facial eczema.

BLOCK 2
Seedling Nursery/Container area.
Young poplar and willow seedlings produced in 1982-83. Note: seedlings derived from crosses of *Populus ciliata* from the Himalayas with the American black poplar *P. deltoides*. From these seedlings improved rust and possum resistant clones will be selected for field trials.

BLOCK 5 – Gully
Opossums are fed foliage of new plant materials selections to determine palatability prior to field testing.

BLOCK 6
Basket willow bed
Five willow clones suitable for basket making planted at various spacings to study cane quality and quantity. *Salix viminalis* 'Gigantea';
S. hippophaifolia (rust susceptible); *S. purpurea* 'Green Dicks'; *S. triandra* 'Black Maul' and *S. Purpurea* 'Nana'.

BLOCK 17
Summer flowering Bee willow *Salix triandra* 'Semperflorens' introduced from Czechoslovakia (origin Tashkent); released 1983.

BLOCK 21, 22
A collection of 90 shrub species maintained to supply seed and cuttings for propagation, e.g., *Atriplex, Ceanothus, Cistus, Eleagnus, Hedysarum, Hippophae, Medicago, Rhus, Symphoricarpos, Tamarix, Teucrium*.

BLOCK 23
Perennial legume trial to determine rates of establishment and ground cover. *Lupinus polyphyllus, Lotus corniculatus, Lotus tenuis, Melilotus officinalis, M. alba, Onobrychis viciifolia*.

BLOCK 24
Legume/grass mixtures: rate of establishment, ground cover and dry matter production.

"If you do erosion control it is better to use plants that have another use than just fixing erosion. Trees should be used for timber or other products, such as foliage for stock or flowers for honey bees. Grasses and legumes should be palatable or productive."

BLOCK 25
Stockbed of Dryland species. *Cistus* (rock rose), *Atriplex* (salt bush), *Sanguisorba* (sheeps burnet), *Dorycnium, Ceanothus, Elaeagnus, Caragana, Kochia* and *Medicago* accessions for seed increase and phenotypic observations.

BLOCK 31
Specimen trees of a wide range of Acacia species. The block contains 50 species.

BLOCK 33, 60
Eucalyptus specimen trees established for observation of form, growth rate and insect damage. Contains 60 species. A further 15 species are planted around the Centre.

BLOCK 37, 38
Shrub and osier willow wand production. These blocks contain clones and hybrids selected for larger scale testing for revegetation and riverbank protection. All are of male sex and are unpalatable to opossums. Includes several single clones and four multiclonal varieties. Only male clones are selected as female clones produce large numbers of seedlings which can cause choking of river channels when used for river training purposes.
These blocks produce each year some 100,000 wands (1 – 1.5m long shoots).

BLOCK 39, 40
Shrub production blocks. Including species of *Symphoricarpos, Cornus, Hippophae, Phormium tenax, Robinia pseudoacacia* (black locust), *Gleditsia triacanthos* (Honey locust), *Alnus* and *Betula* seedling selection and increase block.

BLOCK 42
Biomass production of willows
The current interest in energy farming stimulated the establishment of this trial designed to determine the biomass production of six willow clones at five different spacings and two rotation lengths. Earlier measurements made at the Centre indicated that willows could produce 25-30 tonnes of dry matter/ha/yr on a one or two year short rotation coppicing system and may be capable of even higher production.

BLOCK 45
Seed increase of *Dorycnium hirsutum* and *D. Pentaphyllum* lines.

Dorycnium is a small leguminous shrub promising as a ground cover with fodder potential for semi-arid regions.

BLOCK 56
Shelterbelt trial with various new poplar and willow species and clones to evaluate suitability for shelter.

BLOCK 74
Eucalypt coppice block. Trial to evaluate the potential of coppicing Eucalyptus species for use in shelterbelts. Factors under consideration are species suitability, optimum cutting age, and effect of season of cutting on regrowth.

BLOCK 75
Willow fodder production block. Leaves and shoots harvested daily for feeding trials with penned sheep. (In conjunction with Applied Biochemistry Divison, DSIR).

SHELTERBELTS

Nearly 10 km of shelterbelt has now been planted using some 230 different species and clones.
For further information on shelter species contact: The Scientist in Charge National Plant Materials Centre, Soil Conservation Centre, Aokautere, Ministry of Works and Development, Palmerston North New Zealand. See Permaculture Journal issue 16, May 1984.

WILLOWS

Willows are an important tree for beekeepers as they

are a valuable source of pollen and nectar at a time when many other plants are not flowering. Willows are dioecious – male and female flowers are borne on different trees. The male catkins are the more valuable as they yield both nectar and pollen.

There is a great difference in time of flowering between the various willow species. Mr. van Kraayenoord and scientists at the centre have been studying more than 150 willow clones for several years recording sex, time of flowering and time of flushing.

It is possible for the beekeeper to select a range of willows which can supply pollen and nectar during a long period from mid-July to early November.

The earliest flowering pussy willows such as *Salix medemei (S. aegyptica)* and *S. discolor* yield nectar and pollen from mid-July to mid September, while the weeping willow *(S. babylonica)* and Peking willow *(S. matsudana),* both female trees, are also an early source of nectar.

The male flowers of the crack willow *(S.fragilis)* are probably the most important source of pollen and nectar in September and early October.

In some districts pollen and nectar are in short supply in late October into November and the selection of late flowering male willows such as *S. gatuagensis* and *S. pentandra* will guarantee a continuous supply. The newly imported species S. Semperflorens will extend the period of willow flowering until early April.

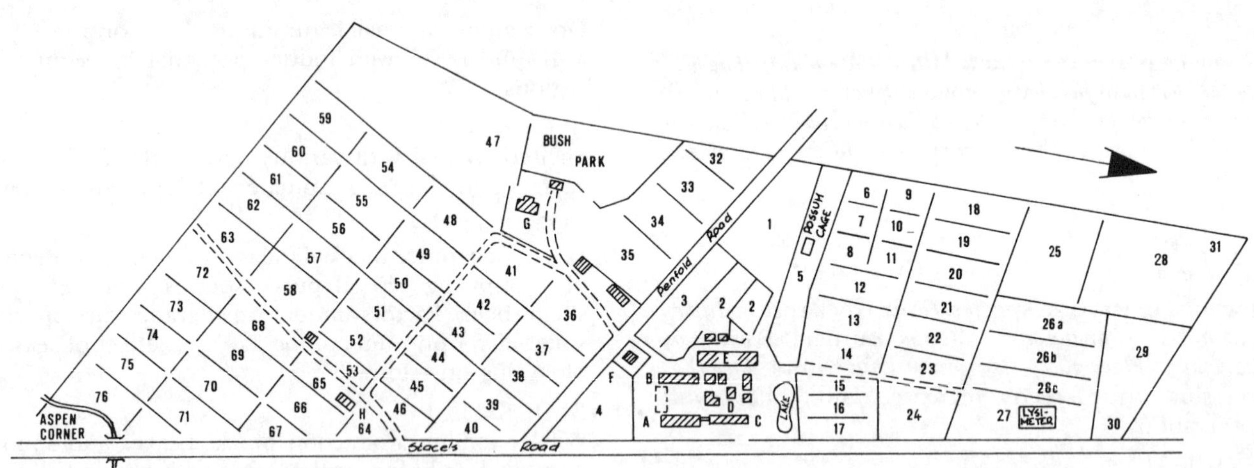

THERE'S MORE TO SOIL THAN MEETS THE EYE

By ALAN SMITH

By burning plants and analysing the remaining ash, the German Chemist Baron Justus von Liebig sought to disclose the so-called "nutrient needs" of plants. In a book entitled *Chemistry In Its Application To Agriculture* published in 1840, Liebig asserted that what plants needed was simply nitrogen, phosphorous and potassium — N,P, and K. At that time the existence of bacteria was not realised.

Now the picture is not quite so simple. Biochemists (that term tells a story!) talk of bacteria, fungi and actinomycetes (often 3000 million per gram of soil) chemicals such as enzymes, amino-acids, antibiotics and gaseous chemicals such as ethylene that are either beneficial or harmful to microbes and plants, and last but not least, the complex organic molecules that are collectively called humus.

1. As plant roots push their way through the soil they exude all manner of organic chemicals – amino acids, sugars, organic acids, vitamins, enzymes and gelatinous materials. These are immediately pounced on by the multitudes of bacteria that are growing along with the root.

2. As much as 30 or 40 per cent of the food reaching the plant roots is lost as exudates of broken cells and devoured by the large microbial populations around the roots. The activity of these microbes dissolves nutrients from minerals in a form that the root can take them in.

3. This is particularly true of the mycorrhizal fungi which sends out many fine food-collecting strands through the finer channels in the soil and takes nutrients back into the root in exchange for food from the root.

4. An added function performed by fungal hyphae is to improve the structure of soils. The hyphae behave like "sticky string bags" in holding soil particles together in aggregates. In a pasture there may be as many as 2 to 5 km of hyphae per gram of soil.

5. Bacteria are really fussy. Each species seems to like only one or two kinds of food. They eat, grow, divide in two, die or if circumstances demand it, go into a sort of resting stage. Actinomycetes-branched bacteria, are very good at breaking up woody residues

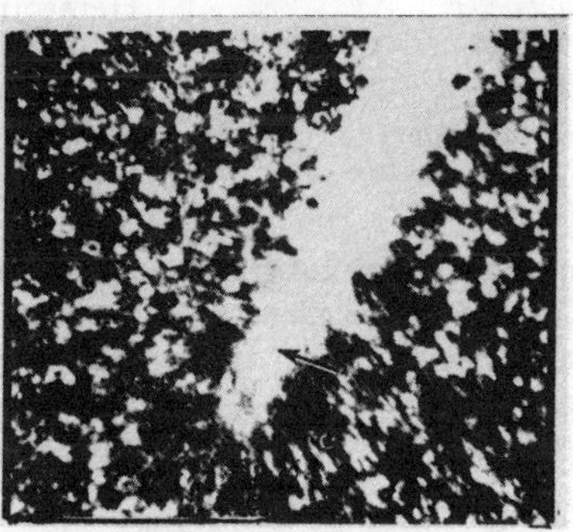

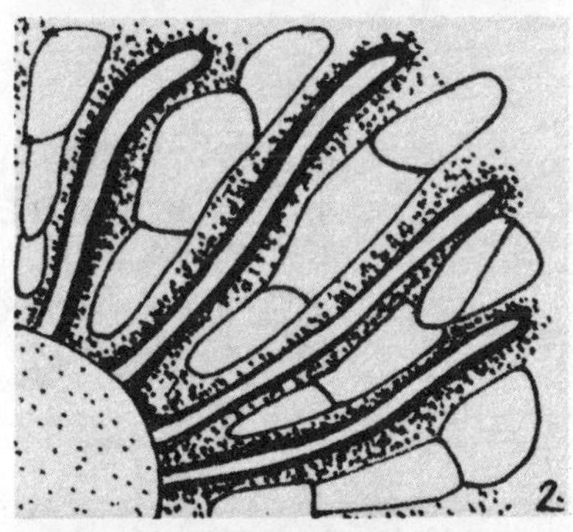

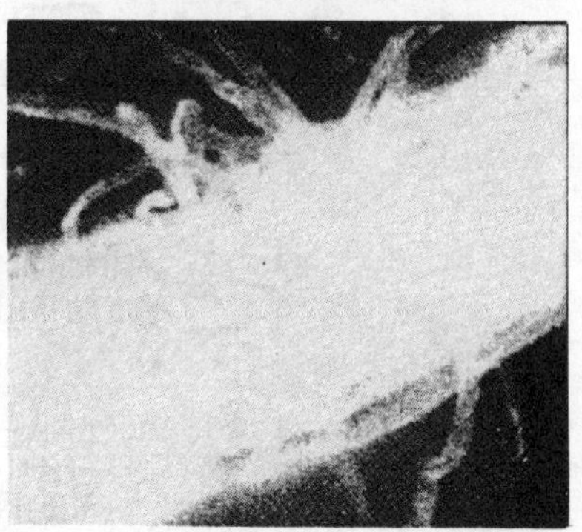

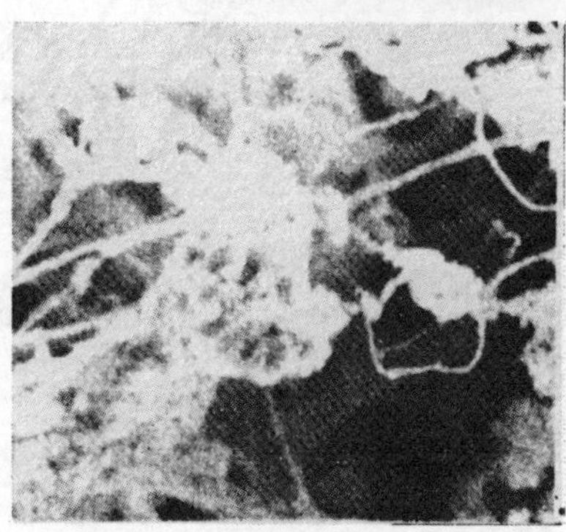

REPRESENTATIVE ORGANISMS	TYPE OF METABOLISM	MATERIAL UTILISED	NAME OF PROCESS
Pseudomonas	Autotroph, Anaerobe	Nitrites and Nitrates	Denitrification
Thiobacillus	Autotroph, Anaerobe	Nitrites and Nitrates	Denitrification
Azotobacter	Heterotroph, Aerobe	Nitrogen Gas	Nitrogen fixation
Clostridium	Heterotroph, Anaerobe	Nitrogen Gas	Nitrogen fixation
Nostac	Photosynthetic, Anaerobe	Nitrogen Gas	Nitrogen fixation
Rhodospirillum	Photosynthetic, Anaerobe	Nitrogen Gas	Nitrogen fixation
Rhizobium	Meterotroph Aerobe (legume association)	Nitrogen Gas	Nitrogen fixation
Many species of fungi, bacteria & actinomycetes	Heterotrophs, Aerobes and Anaerobes	Protein, amino acids and nucleic acid	Ammonification
Nitrosomonas	Autotroph, Aerobe	Ammonia	Nitrification
Nitrosococcus	Autotroph, Aerobe	Ammonia	Nitrification
Nitrobacter	Autotroph, Aerobe	Ammonia	Nitrification

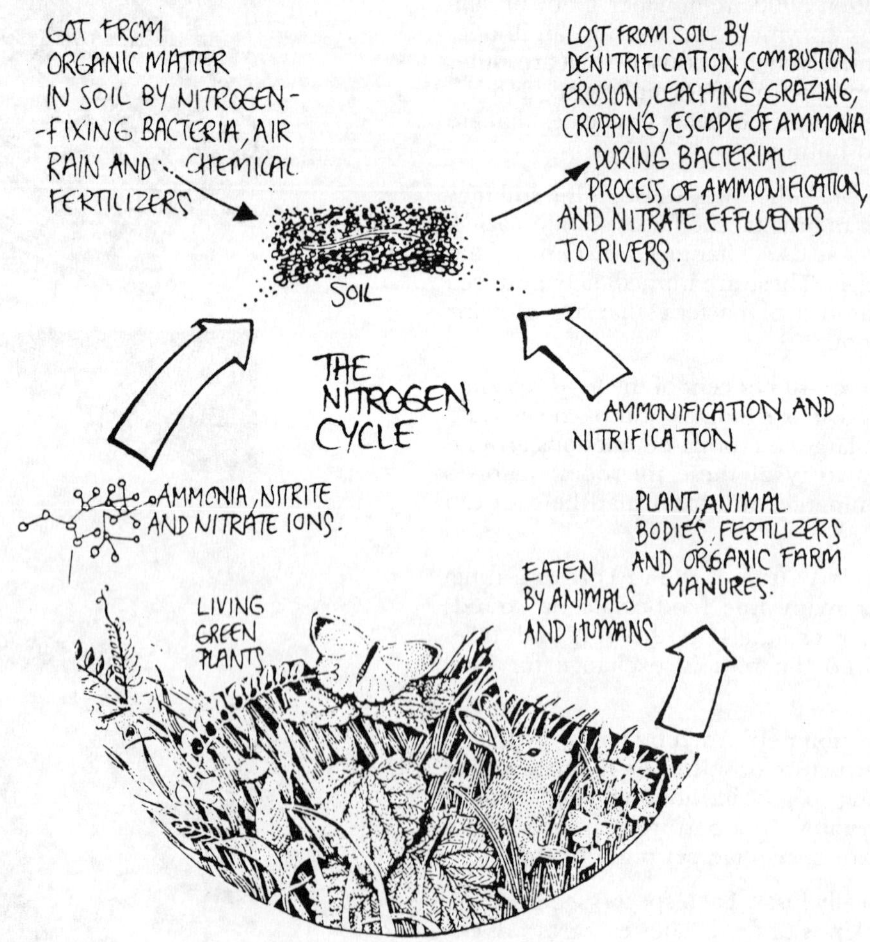

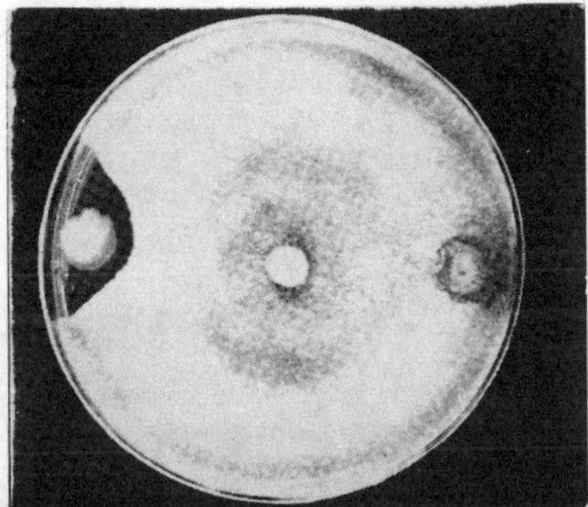

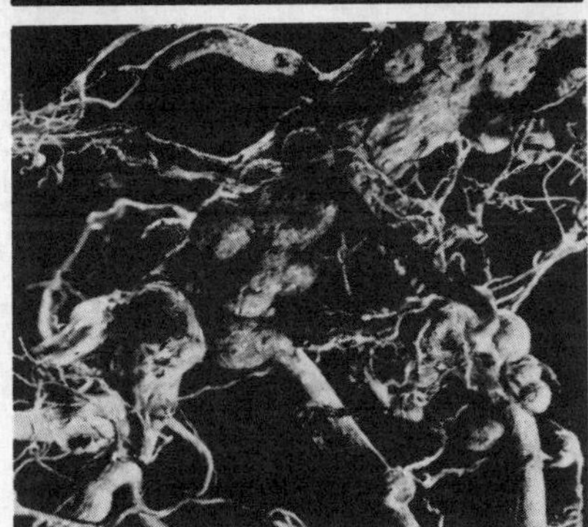

(Photos, Author)

and can survive with smaller amounts of nitrogen than are needed by other bacteria.

6. Of particular importance to the balance of nature is the relationship which exists between many leguminous plants and soil bacteria belonging to the genus rhizobium. The bacteria infect the legume root, and the plant, in reaction, forms swellings (nodules) on the root surface.

7. Within these nodules the rhizobia proliferate and thrive. There they absorb air from the soil and, by processes not fully understood, "fix" the nitrogen. The plant host absorbs much of the nitrogenous product and uses it to produce protein, vitamins, and other nitrogen containing compounds. Cereals, lacking this symbiotic arrangement, have less seed protein and require soil nitrogen or fertilizer for satisfactory growth.

When man introduces a legume to a region where it has never before been grown, the most effective nitrogen-fixing rhizobial strains may be lacking. For the major legume crops, appropriate strains of rhizobium have been identified. The bacteria can be added directly to the soil but usually the legume seed is coated or 'inoculated' with the culture.

Aside from these symbiotic bacteria there are free living ones (see table) which transform up to 40lbs. per acre of the carbohydrates from organic materials to proteins. And some bacteria break down proteins, form ammonia, and then nitrites. Others then form nitrates, and all are used by plants.

Today when scientists dry plants and reduce them by a dehydrating process different from Liebigs simple burning, they find 95% of the dry weight of a plant is material that has been organically synthesised. Half of that is carbon. They also analyse sap as it comes up the stem from the soil solution and find that its mineral content is only .03% to .13% of the total. Many now believe that the actual role of these minerals is to be catalyzers in the organic processes.

REFERENCES

Organic Matter and Soils
C.S.I.R.O.'s Discovering Soils Booklet Number 7. 52 pages $2.00 from Govt. Printers in each state. An excellent reference.

Radical Agriculture
Edited by Richard Merrill 1976
Harper Colophon.

The Organic Gardener
Catherine Osgood Foster 1972
Vintage Books.

The City Forest
P.A. Yeomans 1971

Agriculture
Rudolf Steiner 1958

Tropical Legumes, Resources for the Future
National Academy of Sciences Washington D.C. 1979

DOUBLE MULCHES AND DEEP LITTER

By J. HARRIS

When hens are happy to make us an ideal fertiliser, in the form of deep litter, there is not much sense in labouring ourselves to make compost. And, if the hens will not turn the stuff properly unless it contains plenty of insects, it is better and easier to organise an insect supply than it is to go regularly into the henhouse and turn over everything on the floor with a fork.

A method of insect collecting, which I am just trying, is also a method of litter collecting. It starts with cutting a few 1.5 meter lengths of meter-wide wire netting and weaving bamboo rods through the cut ends. These simple trays are laid down in shady spots and each covered with a double layer of large leaves, such as those of Acanthus (spiky leaved herbs e.g. Sea Holly) or of Chinese paper-leaf, both of which I am delighted to find a use for, at last!

On top of the leaves go path rakings, green weeds, crop residues, rotten fruit, rotten wood hedge clippings and so on. Some manure is sifted in to start the very desirable rot and a handful of fine shell grit added, for the benefit of earthworms. Each small pile can then be covered with an old sack or brushwood to keep the hens out and keep the worms warm at night. When ready which means full of worms and insects, the tray is picked up by the two bamboo rods and tipped out on the hen house floor. Ten or a dozen trays, emptied and refilled in succession at the rate of one every three or four days, should keep the garden tidy, the hens contented and the deep litter in full production.

Another source of insects, in my own garden, is the residues of double-mulches, when raked up after the vegetables have been harvested. In double-mulching, the lower mulch must consist of good earthworm feed. Old stable manure, compost or high-protein chopped green leaves will do, but I usually use poultry deep litter. The upper mulch is of permeable material of low nitrogen/carbon ratio, such as straw, old fern or hedge clippings. Sawdust is unsuitable, being too obstructive to the circulation of air. My own standard material is the debris of dead clumps of pampas-grass.

Protected from sun and birds by the upper mulch,

earthworms can operate 24 hours a day, taking the litter down into the topsoil and soon making it friable. At the same time the upper mulch where it is in contact with the high-nitrogen lower mulch, rots away quite rapidly.

Another angle is that a low nitrogen – or high energy – mulch such as straw can lower crop production by drawing nitrogen from the soil. But a layer of manure, under the straw as in double mulching, prevents this. If instead, and as often recommended, a manure such as blood-and-bone is scattered over the straw, we are likely to get moulds.

The practice of putting quantities of newspaper under mulches, "to suppress the weeds" it is said, greatly impedes air circulation and gains nothing. Last spring I planted pumpkins and choko in small clearings in long grass, scattered sifted deep-litter over the grass, and then packed Eucalyptus foliage thickly over the grass, between the plants. The grass was almost completely suppressed and the plants thrived. Just a little more top mulch would have given all-over grass suppression, if that is thought desirable.

My only mulch failure was with a row of onions. Where deep-litter was the lower mulch between onions and beetroot, the deep-rooting beets were excellent, but the onion plants curled as though struck by hormone drift, their shallow roots finding the soil-solution too strong, I presume. Some means of measuring the strength of deep-litter would be

96

useful. A year before, onions were at their best mulched with leaf rakings from a bush path, with maybe 10% of sifted deep-litter mixed in.

When, if ever, deep-litter output is chemically monitored, the input of kitchen and garden rubbish can be increased or decreased to give the desired level of nutrients in relation to humus in the siftings.

When sifting, I like to have a 10 kilo paper flour-bag handy, to receive bones, corks and odd bits of wood. Corn cobs, which will rot down eventually, can go back into the litter. When almost full, the bag is closed and, when the next fire is being laid, goes at the back of the gate as a paper-bag backlog. Thus a small amount of bone-ash is supplied to the wood-ash kept for general garden use. This ash is one thing that should not go into the henhouse; the litter needs to be kept just on the acid side, so that it will absorb ammonia, when this is produced by decay.

The routine for our hens has been that they turn litter and lay eggs in the morning, being allowed free range in the afternoon. When they started churning up my double mulches, to get at worms and insects below, I had to add a third mulch of brushwood, to keep them out. "Think you're clever do you" said an inner voice. "inventing triple mulch! Take another look and you'll see that all you have invented is typical forest litter."

It is worth thinking about a good mulching system which comes very close to what nature does in a forest. It follows that there should be a high proportion of tree material in the mulches to supply adequate potash, calcium, magnesium and the rest.

For a permanent agriculture, double-mulching has the virtue of giving as much, or more, benefit to the subsequent crop as it does to the current one: it cannot push up crops while it is digesting manure. The reason for this is that nitrogen is withdrawn from the soil, to make protein in the multiplying organisms which are doing the digesting. The result was that the soil bore no crops, during those weeks or months.

The composting answer, traditional in most of Asia, is to do the digesting away from the soil, compost pits or heaps. With adequate compost, crops can be grown twelve months of the year, in suitable climates,

A good ground cover and living mulch.

and with no soil deterioration. Double-mulching is a composting system in which work is reduced by doing much of the digesting, not away from the soil, but above it, in the interface between upper and lower mulch, making full use of mulches which are needed also to control weeds and conserve moisture.

The labour of mulching is much reduced if material is raised right on the spot, as bed edgings and shelter-belts. New machinery may need to be designed, more elaborate and high-powered than my netting trays! It must not be forgotten, though that as production per unit area is raised, and outgoings for fertiliser and so forth lowered, a higher labour input per unit area becomes justified. That the quality of the crop is raised, justifies it further.

INTRODUCING THE "WALLACE" SOIL CONDITIONING UNIT

ANON.

The "Wallace" Soil Conditioning Unit is an implement which improves soil fertility by a unique method of soil aeration developed during 20 years of practical research by a farmer engineer in the Kiewa Valley, Victoria, Australia. It can be used for both pasture improvement and crop preparation.

The soil conditioner will till, aerate, cultivate, loosen, and gently lift soils without turning valuable soil micro-organisms, earthworms, dung beetles, etc; up to be killed by excessive heat and light.

After using the implement, the soil appears slightly ridged with narrow cuts 35 cm apart. Beneath each cut, a triangular air tunnel has been introduced in the soil which promotes nitrogen-fixing bacteria and hence nodules on the roots of legumes.

This aerated soil stays warmer in winter and cooler in summer. Rain penetrates and is absorbed, reducing run off and rapid evaporation to the air. Even freshly cultivated soil is protected from erosion because the stabilising roots of plants are still intact. This combination of air, warmth and moisture provides ideal growth conditions for all beneficial soil organisms.

MAIN ADVANTAGES OF WALLACE SOIL CONDITIONING PROGRAMME OVER CONVENTIONAL AGRICULTURAL PRACTICES.

1. Absolutely no super-phosphate or artificial fertilizers are necessary or desirable.

2. Animals reared on conditioned pasture are healthier.

3. Air in the soil means pasture is warmer in winter, reducing frost damage.,

4. No hay cutting is necessary if stocking rates are kept at optimum levels, because pasture growth continues through the winter (in Victoria, at least).

5. Less fuel use and tractor running costs are incurred, because the land is not worked as often.

6. Tree shelter belts establish quickly in conditioned soil, meaning animals health is improved and crop yields are higher.

Geoff Wallace with the five tyne version of his "S.R.U." (Photos, G. Wallace)

7. Profitability is increased due to much lower costs.

8. Soil conditioning can be applied to any farm with any type of stock with immediate beneficial results.

9. Wallace Soil conditioning can be used on marginal land to advantage. The only limitations are extremely rocky ground or very sandy soils.

10. Stubble breakdown is increased, as stubble is not buried, which inhibits cellulose breakdown and uses large quantities of soil nitrogen.

11. Maximum absorption of early rains is possible—yet land never waterlogs due to improved drainage characteristics.

ACTUAL RESULTS ACHIEVED IN VICTORIA

On the Wallace farm near Kergunyah, Kiewa Valley (36.30 S, 147.10 E) large bullocks are reared without supplementary feeding. One particular animal weighed 3040 lbs (1380 kg). (See Photo)

Aerated soils allow sub-tropical trees and plants to flourish in areas where winter snows are occasionally recorded and frosts are common. The plants include Jacarandas, peanuts, gardinias and wattles, macadamias, avocadoes, mangoes and custard

Custard Apple

Macadamia

apples. The soil temperature of cultivated ground has been measured at up to 9 degrees F, (5 degrees C.) higher than that of non-cultivated ground adjacent to it, in winter.

A striking example of the effect of the soil conditioning programme has been noted in the appearance of two Liquid Amber trees on the property. The trees were identical and were planted on the same day 20 metres apart. One was planted into 'improved' soil and the other into the indigenous red unimproved soil. Eight years later, distinct differences can be seen between the two trees. The one planted into improved soil is approx. 50% larger, and regularly retains its foliage several weeks longer in autumn.

Veterinary visits have not been necessary for many years, except for compulsory Brucellosis injections. Pest infestations are almost non-existent.

Corn, beans and tomatoes planted as late as mid-January still crop profusely, extending the growing season significantly, compared to near neighbours. Absolutely no super-phosphate, artificial fertilizers, pesticides, herbicides or supplementry feeding have been used for 20 years on this farm.

Soil samples taken from areas which have been improved show a rise of up to 1.5 pH units (from 4.5 to 6) and significant increases in available Nitrogen, Phosphorus and Potassium. It is emphasised that nothing whatever has been added, except air.

GENERAL PRINCIPLES FOR USE OF WALLACE SOIL CONDITIONER

1. Never cultivate dry land. This invariably inverts clods of soil.

2. Cultivate approximately on contours, but with slight downhill slope from valleys to ridges. This counteracts the natural tendency of water to flow into the valleys, giving more even distribution of available water.

3. Cultivate at slightly less than walking pace. Too fast will cause some soil inversion.

4. Cultivate at a depth of 4"-6" (100-150mm).

5. Rolling the pasture should be carried out between successive cultivations if there is any chance of the soil being inverted. See details of suitable roller below.

CONDITIONING OF MIXED PASTURE
(including Legumes)

a. Follow general principles outlined above.

b. Cultivate usually once a year after rain has moistened ground and follow-up rains are expected.

c. Cultivation can be carried out more often if compaction through over-stocking has occurred, provided soil is moist.

The Wallace Soil Reconditioning Unit in action.

PREPARATION OF SOIL FOR THE SOWING OF CROP OR PASTURE SEED

NOTE: If the soil is very poor, a programme of soil improvement should be undertaken before attempting to grow crops. This involves using the Wallace Soil Conditioning Unit in conjunction with the growing of deep rooted grasses and legumes over a period of up to three years in extreme cases. Repeated cultivation will accelerate the soil building process, but rolling must take place between chiselling if there is any chance of soil being inverted. The land may be lightly stocked with animals during this soil improvement. This will provide beneficial manuring, eliminate the need for rolling, and provide some productivity as the soil improves.

Sod seeder and Wallace Plough combination.

Preparation for crops can now take place.

a. Replace normal shoes with 'Concorde Wing' shoes.

b. When ground is moist, cultivate following general principles. The 'Concorde Wing' shoes cut the roots of pasture grasses and weeds, causing them to die back.

c. Roll the ground. This ensures that no soil is inverted during the following pass.

d. Two methods of seeding can then be used:

(1) Broadcast the seed on the surface, then cultivate again using 'Concorde Wing' shoes.

(2) Cultivate again using 'Concorde Wing' shoes and simultaneously drop seed into the furrows using a seeder attached above the implement.

In either case, the seed will have germinated and have a well developed root system before the competing weeds re-establish themselves.

Method (1) is most suitable for weed eradication and re-planting of mixed pasture. Method (2) is more suitable if accurate spacing of crop plants is required. In either case, losses to birds are reduced as the seed is obscured by dying grass.

SOME QUESTIONS AND ANSWERS

Q. "But if you're taking something off the land, surely you have to put something back?"

A. If you, for example, graze one animal per acre and it weighs (say) 500lb. more after one year, you're actually taking 500lb. per year off that land, right? Now the weight of top soil, clay and subsoil down to 3 feet depth over that 1 acre is approximately 5000 tons. Therefore we have removed a very small fraction of the nutrient and minerals from that land per year.

By loosening and aerating that topsoil, we're creating the ideal conditions for worms to multiply and loosen the earth progressively further down. This makes available, via deep rooted grasses and herbs, all the minerals and nutrients in that zone. If conditions are suitable, some grass roots go as deep as 10 feet. Tree roots go much deeper. The grasses, weeds and trees actually MINE these minerals – all necessary minerals are present, (in most cases) it's just a matter of making them *available*. The roots bring the nutrients up to the leaves, where they're eaten by animals and deposited on the surface. Dung beetles, if present or introduced, spread the manure and thrive in the healthy environment. No more harrowing.

In conventional farming, the land has usually been compacted by hard hoofed animals or has a hard-pan created just under the surface by continued turning over of the top few inches. The roots of grasses simply can't penetrate down into the soil. They must then be fed by applying supplementary artificial fertilisers.

Q. "But what if I don't have a tractor?"

A. If you haven't a tractor, but your neighbour does have one, why not arrange a barter – the use of the tractor for the use of the S.R.U. He or she will eventually thank you for it.

Q. "How can sub-tropical crops be grown in cold areas?"

A. This is not fully understood yet, but the fact remains that they DO grow! It has been shown, however, that the soil is up to 5 degrees C warmer in winter, compared to non-cultivated areas. This is probably because of the amazing life and activity in the ever-improving topsoil, coupled with the 'blanket' effect of the aeration and increased soil moisture.

Q. "Will my productivity go up?"

A. Probably not. In fact it will probably go down slightly, but profitability will increase enormously! Stocking rates may be slightly less over winter than your neighbours, but you won't have to cut hay. Think how much that alone will save you. Think also how much you'll save by not spreading super-phosphate or chemical fertilisers. Animals grow much larger and healthier on 'Wallace' reconditioned pastures, especially if windbreaks provide shelter from cold winds. These windbreaks establish extremely quickly on 'Wallace' reconditioned land. You may not be selling quite as much, but since costs are so low, profit is maximised. Just work out what you spent on super, chemicals (including drenches) and vet bills last year.

Q. "Why will I have healthier, larger animals?"

A. Users are encouraged to allow a great variety of introduced pasture, and native grasses to thrive together in their paddocks. 'Weeds' are not eliminated, but are allowed to grow in small numbers. The more variety animals have to choose from the more likely they are to remain healthy. (Compare this with your dog or cat eating grass when it feels sick). The animals know instinctively what plant to eat for what ails them. For example, Capeweed has been observed to cure a stomach ailment, bracken patches are used by cows on which to deliver their calves. This apparently minimises infections through the umbillicus. By maintaining their health, animals grow very large, and veterinary bills are mininimised. Drenching is unnecessary, since parasitic worms are kept in check by predators in the healthy soil.

Q. "Why hasn't the Agricultural Department encouraged farmers to use this implement?"

A. The Agricultural Department's charter includes the need to find methods of maximising production per unit area. There is no doubt that more production is achievable if animals and crops are artificially fed (and that includes chemical fertilising). There is increasing evidence that all this artificial stimulation is having a detrimental effect on the quality of animals, crops and most importantly, the soil. The costs of all these imputs is rising every year.

Geoff Wallace is completely happy for the Agricultural Department to monitor results and indeed, they have been invited to do so, but if they started monitoring results from 'Wallace'

'Bill Bullock' weighs in at 3040 lb (1380kg). He grew up on pasture that hasn't had super for 20 years. (Pivot Phosphate Co-op recommends 1 cwt/acre of Superphosphate for grazing properties in Geoff's area.)

reconditioned farms *tomorrow,* it would be along time before any definitive statements could be made as recommendations of the Agricultural Department. Conventional agricultural beliefs tend to block peoples' minds.

Meanwhile 'Wallace' S.R.U. users *know* that their land is improving and even once-skeptical neighbours are starting to realise the benefits to be gained.

DETAILS AND SPECIFICATIONS

All units are made of heavy gauge R.H.S. Steel. The heavy duty tynes are protected against breakage by simple low cost shear bolts. The special steel shoes are engineered to enter the soil at the best angle to save engine horsepower and fuel. The shoes are designed to remain in position by using a simple locking cam bolt placed so that soil abrasion cannot cause bolt wear resulting in loss of shoes. Shoes can be quickly removed for re-shaping or renewal.

Hardened steel 'slippers' are available at very low cost to renew worn shoes without having to replace them. All units are supplied with external wheels which enable the chiselling depth to be kept constant, and also allow transportation along the road by a simple height adjustment.

The recommended roller in the text can be as simple as 2x44 gallon drums filled with concrete on a common axle.

SPECIFICATIONS

	3	5	7	9	11
Approx. Tractor Horsepower recommended.	20–25	35–45	50–60	65–75	80–90
Approx. weight (kg) with wheels	500	600	650	700	750
Width (metres)	1.5	2	2.5	3	3.5
Hectares per hour	.4	.8	1.2	1.6	2

The minimum size 3 tyne unit is easily usable with a Grey Ferguson Tractor, or similar.

COMMENT

If seeing is believing, I suggest you mark next Easter on your calendar and attend the Organic School which is held every year on Wallace's beautiful farm at Kergunyah via Wodonga.

If anything, it may just look too easy; the lush growing field of oats, the depth of root penetration, the excellent crop of water-melon and the healthy looking animals.

Looking at properties up and down the valley, the difference can easily be seen. Since Geoff hasn't used any fertilizer for more than 20 years, the answer must lay in the Wallace S.R.U.

Our limited experience under sub-tropical conditions shows that this implement is likely to be of great value in most areas. It is not a soil-breaker, it is a a soil-conditioner, and if used appropriately is excellent.

COMPOST VEGETABLE GROWING

By KIM CHRISTIE

I have been corresponding with Jim Rankin, Director, Scientific Gardening at Solusi College in Zimbabwe. Mr. Rankin is a successful and experienced organic gardener who went to Zimbabwe to train farmers in better growing methods.

When Jim went to the college all needed nutrients were supplied to the sandy soil with chemical fertilizers. Convinced he could show local farmers how to do better using locally available plant materials, he established ten test beds, proved the method worked and changed the project to the composting method. As the normal 3-6 month compost method is too slow, he has developed a 14 day method.

Zimbabwe is one of the worst-hit countries in Africa's drought and Solusi College is really in the hot zone with temperatures regularly up to 46 degrees C. Despite this, on a 2.2 acre plot he produced in one year 98 tons of saleable vegetables with a market value of $23,000. Costs are low and the profit is used to train subsistence farmers free of charge, these graduates then train fellow villagers.

Jim has been very generous and has provided me with his lecture notes and minute detail of everything he does for use in the courses Max and I are running.

I have included just a few extracts from the information he has so far made available to me.

Results of experiments at Solusi College

YIELDS IN Kg

	Fertilizer	Compost	Both
Tomato	150	128	210
Cabbage	234	202	350
Onion	120	118	145
Potato	80	107	128

"This proves to us that if compost is used regularly, yields are almost equal to a good fertilizer programme and, if used in conjunction with legume green manure crops, can be higher than just fertilizer or compost alone."

"Sawdust turned into compost when we put 3 kg Urea into 200 litres of water and saturated 1 cubic metre of dry sawdust with it, then turned it and kept it moistened to 40-60%

1. Compost is applied at 5 kg/sq.m. with dolomite to correct pH as required and worked in to a depth of 30 centimetres.

2. When plants are 12 cm tall, a 2 cm mulch of compost is applied but not worked in. This is watered taking all nutrient ions in solution down to the root zone, the fibre breaks down within 30 days when another mulch is placed on top of the soil taking care not to place it against plant stems. By now leaf coverage prevents rapid breakdown.

3. In their almost pure sandy soil, poultry manure in solution is fed every 3 weeks – 20 litres manure in burlap bag in 200 litres of water. At the start of his programme he used 1 kg Urea in 200 litres of water. If phosphates are very low rock phosphate should be applied.

4. Jim says "We should learn to work with the soil not constantly struggle with it and though I try to use only organic methods in my own home, I do suggest to students that if the soil is very low in one of the 13 essential plant nutrients, the organic matter in the area too will be low, then cheat a little and use nutrients from store bought fertilizers, but only to get the nutrients into the soil and then use: buckwheat as a green manure to get phosphates; tobacco to get extra potassium, and dig them in when the plants are 45cm tall."

NOTE

It may not be necessary to cheat with store bought chemicals if ground rock minerals are used.

COMPOSTING PERTH'S REFUSE

By PERMACULTURE ASSOCIATION OF W.A.

Speaker at our January meeting was Mr. Amos Machlin, City Engineer for City of Perth. The meeting was very well attended and our 'guests' included one or two people who didn't see eye to eye with Amos in his enthusiasm for large scale composting, and the debate became quite spirited sometimes.

Amos opened by telling about the high cost of rubbish disposal – the 80,000 ratepayers in City of Perth Municipality area pay about $3.3M each year toward the rubbish disposal bill out of the City's $20M total revenue, and this is far short of the total actual disposal cost.

So the City decided that they had to reduce this disposal cost by recognising rubbish as a renewable resource, useful to others. A review of overseas methods was undertaken and it was found that in many places, in Europe especially, refuse could be burnt in sophisticated incinerators and the heat generated could be put to good use in providing electricity and comfort and industrial heating. But Perth's climate and industry and scale of operation didn't favour this option. We have a particular problem with our sandy soil and poor water supply. So composting was chosen as having great potential to upgrade the soil fertility in the City's own parks and gardens (and to bring in cash from sale to the public) and also to help reduce Perth's horrific water consumption (and consequent power consumption) rates.

Amos then went on to outline some details of the machinery and process for converting garbage to compost. Their plant will have two big composting tumblers – each one holds about 200 tonnes of material and is 3 metres in diameter and 40 metres long, turning constantly at about 3 revolutions per minute. The raw untreated rubbish is fed into the drums and is retained in that first part of the process for 2 or 3 days. The fine organic material is then separated from the coarse inorganics by screening, with the coarse non-composting part going for disposal by incineration or landfill. The fine material is treated further, going to large stockpiles for 2-3

weeks where it heats up enough to kill off any pathogens (bad bugs), then to composting piles for 8-13 weeks, when it is properly composted and ready for sale. Sewage sludge is added to control the moisture, to help the process and improve the compost. The resulting compost has between 1% and 5% nitrogen depending on the amount of sewage sludge used, and has been given a clean bill of health by European and World Health organisations. Even the W.A. Health Department have approved it!

Problems in the process are, too much paper and too much glass in the raw rubbish, (compost isn't nice to handle full of glass splinters.) So separate disposal of these is desirable, and the City have had encouraging results from a pilot program of having householders separate out their garbage into 4 containers – one each for cans, glass, paper and other. Public response has been very good (up to 90% of households doing the right thing) – resource recovery can work, and valuable tip space is being saved.

COMMENT

Perth's solution to the waste problem is not unique. It's happening in many places around the world. Rubbish doesn't make headlines very often. It is up to all of us to make our councils aware of possible solutions. After all it is our resources that are wasted!

THE JEAN PAIN METHOD

By BRUCE FULFORD

In 1981 Bruce attended a two week workshop in Belgium on the work of Jean Pain, pioneer French researcher into biological energy usage.

A two week workshop on the brushwood energy and composting methods of French researcher Jean Pain was attended by 35 participants from Africa, Europe, Great Britain, and North and South America. The six representatives of African nations were funded by the U.N.'s Food and Agriculture Organization.

The Jean Pain method uses green brush and tree prunings (chipped, soaked, and stacked in piles weighing up to 200 tons) to generate hot water, methane, and create soil in large quantities where there are only trees or brush to provide organic matter.

Pain developed his methods after a lifetime of observation of the natural world. He saw man's role as either destroyer or preserver of the planet and that by learning from and cooperating with nature, we can live in a permanently balanced and healthy lifestyle.

Jean Pain lived in Provence, France, a hot and dry Mediterranean climate, where thousands of years of poor soil conservation left little soil and forests of scrub pine and brush, which became tinder for devastating fires almost every summer. He discovered that by thinning the small branches and brush before they became dry and flammable he reduced the danger of fire and that he could compost this brushwood to rebuild the lost soil that is essential to a healthy environment.

Pain also developed methods of extracting thermal energy and biogas (methane) from this green brushwood, providing gas for cooking and transportation and hot water for house heat and bath water.

The workshop was organized by the Comite Jean Pain and was taught by Jean Pain's nephew, Etienne Bonvalet. The workshop covered all aspects of the theory and practice of producing compost energy from green brushwood, including forest management planning, equipment and labour requirements, payback periods, gas and hot water production, finished compost value, and applications for industry and agriculture.

Tractor driven Jean Pain shredder. (Photo, Neil Druce)

A tour of several sites in Brussels using this method, exhibited some of the problems and solutions encountered in the course of experimentation in a variety of situations. The classroom sessions of the workshop were held at the University of Brussels and the practical sessions were held every afternoon in Londerzeel, where the research site and gardens are located.

Perhaps the most remarkable aspect of Jean Pain's method is the fact that these piles of chipped brush attain temperatures of 140-150 degrees with no additional nitrogen source, which most composters have come to believe necessary to break down concentrated lignins in wood fibres.

What makes this possible is the use of green brushwood chipped, or preferably slivered, to a thickness of about an eighth of an inch, which is then soaked for 24 to 48 hours and assembled in piles no more than three metres in height. The coarseness of the chips provides a pile structure that permits good composting without forced aeration.

The maximum thickness of a quarter inch allows for penetration of the water and air necessary for bacteria to biodegrade the woodchips. The maximum height of three metres assures the airspaces do not become too compacted, resulting in the lower portion of the pile becoming anaerobic.

For heat exchange, Pain's research at the test site in Londerzeel, Belgium showed high-density PVC tubing to be superior to copper or other materials tested in cost effectiveness and overall performance.

PVC is far less expensive than copper, and, though copper conducts heat better, the physical nature of compost is such that if heat is rapidly "wicked" out of the area immediately surrounding the pipe, the ambient temperature of the pile cannot be passed to the water travelling through the pipe. Copper has also proved to be more susceptible to chemical or bacterial degradation thanPVC.

PVC is easily punctured by pitchforks, so caution must be taken during the pile disassembly process. However, if the tubing is punctured, the holes are easily "soldered" with a special teflon coated "soldering iron" that melts the plastic without sticking to it. The process is used to join segments together and to eliminate the many joints and sleeved unions with hose clamps that would rust in the compost pile. Experimental 'spot welded' tubes hooked in series were installed at the workshop. These can slide out of the pile easily prior to disassembly, to elimate the danger of puncture. Hose clamps are located on the outside of the pile.

Several different "broyeaurs" (chippers) were used to chip the brush. The brush had been brought to the site by local landscapers who do not have to pay a fee to dump trees and brush at this site, as they would at the dump. Some brush was also cut from the privet hedges, willows and poplars - all fast growing "coppicing" trees at the site that provide large amounts of biomass rapidly.

One major point of Jean Pain's method is that the finished compost is a balanced, complete food that is ideal for plants. The product tests out to be lower in nitrogen than many composts using manures and high nitrogen solutions, but the plants grown in this lower nitrogen compost are of a perfectly balanced composition.

The gardens in Londerzeel seem incredibly pest free, healthy and productive. This is a dramatic change from the hard-packed poor soil that was there five years ago.

Because the workshop was attended by people who live in a wide range of climates - Austria to the deserts of Senegal, to the jungles of equatorial Africa - these methods need to be adapted for each particular climatic situation. A joint project is now underway in Tunisia with the help of the Belgian government and Comite Jean Pain to set up a composting operation of this method. It is hoped that the resulting compost will provide the basis for expanding the green-belt region. The reforestation of the desert through the creation of soil by composting brushwood may be one solution to the spreading famine created by the desertification of productive lands.

The book *Methods of Jean Pain – Another Kind of Garden* has sold more than 70,000 copies without advertising. The proceeds are used to fund further research, workshops and to provide more literature. Pain's widow, Ida, plans to continue the work and publication of the book, now printed in five languages.

Jean Pain died July 30, 1981, the last day of the workshop. He was fifty years of age. Pain said that he felt that this year there would be an "explosion" of the methods which he had developed in the last 17 years. We deeply mourn the loss of an inspirational force and person.

The Comite Jean Pain receives mail daily from all over the world asking about this method of balancing agriculture, energy and forest management. The BioThermal Energy Centre also provides information on this method. We carry Jean Pain's book which sells for $12 postpaid. The Centre will offer slide shows, films, and speaking engagements on Jean Pain's methods and other aspects of BioThermal Energy in the future. Please write for more information.

* Bruce Fulford is coordinator of the Biothermal Energy Centre, Portland, Maine, U.S.A.

COMMENT

The Jean Pain Method is very well suited for all areas which have a plentiful supply of excess shrubbery. (This would exclude most arid areas.) It should make it possible to turn a firehazard or a weed into a valuable resource.

CONTROL OF FUNGUS DISEASES

By KIM CHRISTIE

There are many references now to the control of insect pests with alternatives to chemical sprays, but I have always had great difficulty in locating any alternatives for the control of fungus diseases which are so prevalent in the summer rainfall areas, and seem to prevent most people from growing a continuous supply of summer vegetables.

During the Permaculture Course in Tasmania, I found a reference to the use of Horsetail *(Equisetum arvense)* tea and have since researched this further and used it during this summer with very good results.

Every summer I have lost virtually all lettuce, celery and parsley with sclerotinia rot, zucchinis and cucurbits with mildew and tomatoes with various rots, beans with rust, etc. This year there have been no losses at all in lettuce and celery, a few tomato losses, none in beans, and zucchinis have borne continuously since October without any sign of mildew.

Equisetum contains 80%-90% of Silicic Acid and the silica appears to strengthen plant cell walls preventing their invasion by fungus spores, as well as having an inhibiting effect on the spores themselves. It appears to be effective against mildews, rust, scab and sclerotinia rot as well as anthracnose and peach curly leaf. *Equisetum* tea is not a cure once the disease is present but has a preventative action, it can be sprayed from the time the first true leaves mature and should be applied fortnightly in wet or hot, humid weather.

The dried herb can be obtained from health food shops. I have written to the major sellers of herb teas to try and obtain a cheaper bulk supply but without even a reply. *Equisetum arvense* apparently is not grown in Australia. I have recently managed to contact a couple of large commercial bio-dynamic farmers in New South Wales and been advised that casuarina *(Casuarinaceae* Family) needles contain almost the same proportion of silicic acid and are used in the same way.

To make the spray:

20 g. dried leaves (80 g. green casuarina needles)
1 litre water.

EITHER boil for 20-30 minutes, cool, strain and spray OR bring to boil and allow to ferment for 2-3 weeks, strain and spray.

I usually add Seasol (a liquid seaweed extract) to the spray.

As well as spraying plants, seed beds and boxes, glasshouses, etc. it can be sprayed to reduce fungus attacks, and the ground under fruit trees can be sprayed early spring and again in Autumn. Spray need not be heavy as it appears to work homeopathically; one litre can be used to cover up to 1,000 sq. ft. After the first couple of sprays have been used, the dose can be diluted in successive sprays by 50%.

For fruit trees a paste can be made to paint onto trunks and lower limbs which destroys overwintering insects, insect eggs or spores.

I have found two recipes for this:
1. 2 kg clay
    1 kg cow manure
    3 *l* equisetum tea.

2. 1 kg clay
    1 kg diatomaceous earth
    1 kg cow manure
    equisetum tea to make paste

Seasol or other seaweed preparations can be added to both.

If wished the mixture can be further diluted, strained through a cloth and sprayed on if there are a lot of trees to be treated.

Other alternative fungicides are wettable sulphur, sulphur dust, Bordeaux, copper oxychloride, Condy's Crystals. However, be careful as sulphur can burn some crops.

For powdery mildew 7 gm. of Condy's Crystals dissolved in 7 litres of water and sprayed at once is effective.

For downy mildew:
112 g. washing soda
 56 g. soft soap
 5.5 *l* water

Both these are completely safe.

There is some objection to the use of copper sprays but a high humus content soil will act as an effective buffer against copper toxicity.

Remember, heavy feeding (organic or inorganic) will cause cell walls to weaken and cancel the beneficial effect of the sprays mentioned.

REFERENCES

Koepf, Petterson and Schaumann – *Bio-Dynamic Agriculture,* The Anthroposophic Press, New York.

Steiner, Rudolf – *Agriculture,* Bio-Dynamic Agricultural Assocn., London.

Philbrick and Gregg – *Companion Plants and how to use them,* The Devin-Adair Company, Connecticut.

Brisbane Organic Growers – Pesticides and Alternatives, Vibro Press, Brisbane.

BIOLOGICAL CONTROL IN PERSPECTIVE

From "ECOS", produced by the Scientific Communication Unit of C.S.I.R.O., Canberra Act.

Praying mantis ona guava tree. Note the grass hopper to the right.

Dealing with plant or animal pests using biological control has many advantages over using pesticides or mechanical means, of which the most important are that:

– Once it has succeeded it is more or less permanent, so it's cheaper in the long run

– it affects only the target pest

– it doesn't pollute the environment

However, no doubt because of its more spectacular successes, the method is the subject of a number of popular misconceptions.

Perhaps the most commonly held of these is that biological control is an alternative means of eradicating pests. It isn't. The method does not seek to totally annihilate weeds or animal pests, it aims at permanently reducing their numbers to a level that we can tolerate. Thus prickly pear still grows in Queensland despite the fact that the little moth Cactoblastis cactorum very successfully controlled it nearly 50 years ago. But nowadays the cactus rarely becomes plentiful enough to be regarded as a pest. Small populations of the moth remain in the pockets of prickly pear that survive, ready to attack any flare-ups.

If successful control agents can be found, this situation should apply equally well with water weeds like water hyacinth, salvinia, or alligator weed, or with any other plant or animal pest. Thus, in the long run, biological control becomes by far the most economical method.

Another major misconception is that biological control will always be quick. Biological control is usually remembered for the spectacular speed of its successes – it took only a few seasons for prickly pear and the most common strain of skeleton-weed to be controlled after the release of suitable control agents. Usually the results will be much slower in coming. A period of several or even many years may elapse before the introduced control organisms have increased sufficiently in numbers and range to begin controlling infestations of a pest.

The philosophy behind biological control is of course, simple. In their natural state plants (and animals) occur in population densities that remain in balance with the rest of the biological community. Gross disturbance of that community may upset the balance, with the result that some plants become extinct, while others become most dominating.

Agriculture causes just such a disturbance, and the successful but unwanted plants that grow and compete with the crops or pasture plants are, of course, weeds. The most extreme case of disturbance occurs when a fast-reproducing plant or animal is removed from its place of origin and introduced to another continent.

Under these circumstances the biological checks that limited this organism are completely removed and it will rapidly multiply wherever circumstances permit.

The Australian landscape is littered with examples of exotic plants that have been introduced and then become weeds – prickly pear, skeleton-weed, lantana, gorse, and Paterson's curse (salvation Jane) are but a few examples, and the explosion of rabbit numbers in Australia last century showed very clearly how an animal can become a pest once constraints on its breeding have been removed.

Such pests seldom cause problems in their countries of origin, so something, or more often several things, are keeping them in check. If some of these checks are living organisms (like insects, or diseases caused by fungi or viruses) it may be possible to use them to control the pest in the country where it has become a nuisance.

This may sound simple, but an effective biological control program involves considerable amounts of research. To begin with it is essential that the controlling agent should confine its attention to the target pest, and also essential to make sure that it will not attack crops or other valuable plants. Also, an organism that attacks the particular pest in its place of origin won't necessarily do so effectively in its new home. The environmental conditions may be different in the two places, and although the pest may thrive, its introduced enemies may not. Thus careful evaluation is needed to make sure that the introduced organism is suited to the environmental conditions of the new country. Even organisms that have successfully controlled weeds in one country after introduction there may not succeed under the different conditions prevailing in another – like Australia. (Sometimes the converse may apply. An insect that has little effect on the target pest in its place of origin may thrive on the weed away from its own enemies in the new country.)

As may be expected, introduced control organisms are usually more effective if they can breed quickly,

Lady beetle population on the increase in response to this aphid infestation.

and distribute themselves rapidly. By the time a plant has become a troublesome weed, infestations will nearly always be covering large areas. Control organisms are therefore released in as many localities as possible, and left to spread themselves.

In summary: organisms that confine their attacks to the target pest, breed rapidly, and originate from an environment similar to the one into which they will be released have a good chance of being successful biological control agents.

GROUNDSEL UP-DATE
(Baccharis halimilfolia L.)

By Max O. LINDEGGER

BACCHARIS L. COMPOSITAE 400 spp. America, Herbs or Shrubs.

A dense shrubby perennial with toothed leaves, yellow or white flowers and 'seeds' like thistledown.

Flowering time:	March – April
Colour of honey:	Light Amber
Importance as source of honey:	Minor
Importance as source of pollen:	Medium
General Remarks:	This bush provides a valuable pollen supply in coastal districts during autumn and is worked heavily by bees.

(Blake and Roff, *The Honey Flora at Queensland.* D.P.I., 1972)

After many years of being on the Noxious Weed List and attempts at eradication by chemical and other methods, the spread of Groundsel has, at best, been held in check. No area, to our knowledge, has reported the eradication of the weed, but many areas show recent infestation.

The cost to local authorities is very high and to landholders very much higher. With the very limited success chemical methods can claim, and the increased opposition to the use of hormonal herbicides, alternative controls are constantly assessed.

The Alan Fletcher Research Station (QLD) has been searching for biological agents for a considerable time. Before being introduced, the insects are subjected to years of careful and detailed testing to ensure that they affect only Groundsel bush and will survive under the conditions they will be exposed to. Two insects, the Stem Boring Beetle and the Gall Fly have been encouraging. In the test area it has been shown to reduce flowering by 80-90%. Gall Fly can reproduce every 4-6 weeks during ideal conditions which exist in early and late summer. It is expected that it will take 1-2 years for this biological agent to spread to effective numbers.

The Gall Fly *(Rhopalomyia Californica)* has attacked all flowering heads on this Groundsel plant.

With the availability of the Gall Fly at least two options are open to deal with Groundsel:

1. HERBICIDES

Herbicides are still the cheapest short term method. However, experience shows that herbicides like 2,4-D are not a permananent solution. 2,4-D will not only kill Groundsel, it will also affect other, often very desirable plants (eg. rainforest trees, wattles, hoop and bunya pines, Eucalypts). In many cases where Groundsel had been killed by 2, 4-D the land became infested with wild raspberry and blackberry. The Department of Primary Industry and Local Council recommended weedicide for these plants is 2,4,5-T.

The evidence against the use of chemical sprays and the increased pressure from people affected physically and psychologically from having to spray or otherwise deal with Groundsel has led to our suggestion that the following option be considered.

2. RE-FORESTATION, PASTURE IMPROVEMENT

- Maintenance slashing has in isolated cases eradicated Groundsel (but generally will only prevent it from flowering). Inaccessible areas should be cut by hand.

- Biological agents should be released in concentrated numbers at the time of slashing or shortly following.

110

- Ripping where possible, will aerate the soil and increase water absorption.

- Forest trees planted at spacings not greater than 4m will shade the ground. Groundsel does not thrive under such conditions, and a 80-90% control can be expected within 3-5 years in most cases.

- Maintenance slashing will be necessary during this period.

Pasture improvement involving ripping, seeding of suitable grasses and legumes and necessary fertilizing combined with sensible stocking rates, has also shown to create conditions unsuitable for Groundsel.

CONCLUSIONS

While the above article is specific to one weed, Groundsel, one will find that the methods used to deal with it can be applied to many other weeds.

As humans, we have a tendency to deal with very short time spans. If we look at the example 'Groundsel' over a period of one year only, we will find that the chemical 'control' is cheaper than the re-forestation option. If we calculate the cost over a longer period we will find that the 'forest' option is not only the more healthy, aesthetically pleasing way to deal with Groundsel, but is actually quite profitable - and permanent.

Foot Note: Advice on suitable forest trees is available in most areas. Genuine forest schemes attract tax incentives.

COMMENT

The success of the gall fly has exceeded most expectations. While the flowering was prolific in the past season it will be interesting to see if gall fly affected plants will be able to produce viable flowers.

Groundsel has a spindly, weakened form in shaded forest.

Legumes can eventually 'smother' the Groundsel plant, as they compete for nutrients and sunlight.

ORGANIC GROWING OF GRAPE VINES

By GIL WAHLQUIST

At Botobolar we are often asked how we grow our grapevines and in what way our growing practices differ from those of others.

This is an attempt to set down as briefly as possible, how we grow the vines. The article is directed specifically at people who are already growing vines or who may be thinking of doing so.

We start with a philosophy, deciding to go with nature, not against it. To be patient and observant. To minimise cultivation and spraying and to keep a strict control on the type of sprays used.

The important thing to grasp is that organic growing isn't a do nothing system. It calls for a great deal of management and the development of individual strategies for the problems which arise. This preamble is necessary because I can only give information on the things which work at Botobolar. This does not represent a success formula for elsewhere, but it may help a grower develop strategies of his or her own.

Grapevines are easy to grow, provided a location is found which has a well drained soil with water equivalent to 660mm rainfall per year. If regular cropping is desired, spring frosts must be at a minimum.

The chosen area is cultivated to remove weed growth and kept clear for a few months before planting, between June and late August.

Once the planting distances are decided, the rows are marked out and deep ripped. A bulldozer is brought in for the job. The purpose of this is to break through any cultivation pans which have built up in the paddock, assuming that the vineyard is being planted in cleared land with some history of agricultural use.

A study needs to be made of existing wildlife, including rabbits, hares and wallabies. If they have been regular users of the land, fences will have to be erected to keep them out. The locals will probably know the best sort of fence. It will probably be electric. You will need to keep wildlife out until the vines get up to the trellis wire. Remember, the animals were there first.

BOTOBOLAR
SHIRAZ

Once the vines are established, we allow grass to grow in the vineyard area and in the second or third year, depending on the size of the vine, we will allow sheep to graze in the vineyard when the vines are dormant. We did cultivate strips of the vineyard over the years, but following advice from a soil scientist whom we respect, we intend to discontinue cultivation and have purchased a sturdy mower to keep grass growth down. The channels created in the soil by grass and earthworms are more efficient conveyers of moisture than a cultivated situation. Continued cultivation destroys the crumb structure of a good soil so that it tends to set when wet, being a poor transmitter of moisture to the depths.

When it comes to choosing tractors, we prefer a small one. For earthmoving you can always contract a neighbour or district farmer for so much an hour. Your own tractor need be powerful enough to power your spray unit and slasher, but no more.

Vines in their first two years need to be kept clear of weeds around the base for a radius of about 25 cm. Purchase strong chipping hoes and do the job by hand. For a larger vineyard you will need an undervine plough attached to the tractor. These need a 40 hp tractor to operate. You will have to consult the local machinery dealer.

Undervine mowing. (Photo, Botobolar).

Vines are subject to two mildews, downy and powdery. Young vines are susceptible to sprays on the market. We use good old fashioned bordeaux mixture once the shoots become 14cm long. We re-spray every 14 days until the growth is complete, then watch the weather. For Bordeaux you need a spray tank with an agitator. It is made by dissolving copper sulphate crystals in water, then adding hydrated lime which has been made into a milk in a bucket. The two chemicals are used in equal quantities by weight. We use about 4 kg to the hectare of the mixture. We put about 70ml of wetting agent per 100 litres of solution. The wetting agent is essential to get the stuff to spread. Bordeaux mixture needs to go on fairly wet. We apply 250*l* to 300*l* per acre on mature vines in a dry land situation (that is, not irrigated). If the vine leaves have a slightly blue tinge you know you have got the coverage. The chemical needs to get under the leaves, for that is where the downy mildew spores enter the leaves.

Do you spray when the vines are flowering? We have learned not to. Spray seems to interfere with setting of the fruit. You need to spray the moment flowering is complete to get a coverage of the young bunch. Bordeaux mixture has a toughening effect on the leaves and seems to give some resistance to powdery mildew.

For spraying we use a Solo backpack and a Silvan airblast behind the tractor.

Powdery mildew is a problem in cool, dry situations –

in the ideal grape growing conditions. It can sneak up without you being aware of it. Downy mildew lives in the ground. Powdery lives on the vine and spores over winter in the bud scales. So you need to do something about it towards the end of winter. When the mean daily temperature reaches 15 degrees C we spray with lime sulphur. This smells like rotten eggs so it is a job to be done when you are standing well upwind. You work out mean daily temperature by taking the daily minimum and the daily maximum (daily being over 24 hours) and averaging them. To do this you will need a maximum and a minimum thermometer, installed outside somewhere in the shade. We built a Stevenson screen, similar to the ones used at weather stations, out of four louvered cupboard doors and an insulated roof. It stands outside near the vines. Say the minimum temperature is 8 and the maximum 22, the total is 30, divided by two, average 15 – spray the lime sulphur, provided this temperature pattern looks permanent. Why 15 degrees? Well, at 15 degrees the vine starts to grow. There is movement in the buds and the mites, if any, move about.

The lime sulphur will take care of the majority of mites, which are small insects which suck sap. At 20 degrees, the predators on the mites emerge. So, if you wait until after 20 degrees daily average has arrived, your spray will destroy the goodies as well as the baddies.

Powdery mildew can be seen in several different ways. If it is on the leaves it is a grey powdery mould on the top of the leaves to start with, finally browning and killing them. In the grape bunch, it is seen as a grey mould which causes the berries to harden and split. On the vine shoots it causes a drying of the growing tip. If it appears in the vineyard despite the lime sulphur and bordeaux, we suggest the French method of control which is simply sulphur. This is powdered sulphur which we dust on.

We found that we need about 30 kg per hectare on mature vines. This is way over the Australian recommendations but is in line with the French recommendations. Powdered sulphur can burn leaves and fruit in hot inland situations, that is probably why Australian recommended rates are so low. In a powdery mildew situation, the French recommend 2 kg to 6 kg of wettable sulphur at budburst, 15 to 20 kg of dust per hectare when shoots are showing three or four leaves, 5 to 8 kg of wettable sulphur per hectare pre flowering, 20 to 30 kg of dust at flowering and 30 to 35 kg of dust pre verasion, that is when red grapes are colouring or when white grapes are getting that waxy look. Powdery mildew is easy to control by using sulphur. Lack of control can lead to devastating effects on crop and vine. We use a duster which we purchased from Pattersons in Mildura.

There is one other spray that we use and that is winter oil. We only put this out when there is a large infestation of scale. This is a small, sap-sucking insect, usually controlled by predators. The oil needs to be sprayed in winter. It acts by suffocating the scale when they are on the move. It helps reduce numbers so that the predators can cope. We use the rates recommended on the container.

We don't spray for caterpillars. That is where the grass in the vineyard helps. Besides preserving soil structure it acts as a home for a variety of insects, many of them predators on the vine caterpillar, cabbage moth and apple moth, all of which have been known to attack grapes. If you have a new vineyard in an area with a history of spraying, your predators may be insufficient at first to cope with a large infestation of caterpillars. In this case spray Dipel, which is the trade name of a product called *bacillus thuringiensis.* This is sprayed onto the leaves. When the caterpillars eat it, it upsets the pH of their stomach and they stop feeding, eventually withering and dying. The bacillus does not affect the predators on the caterpillars. Dipel has been around for years but has not been popular until recently because it does not have an instant effect. You spray and the caterpillars keep on eating. But experience shows that they soon stop and in the long term your predators take care of the caterpillars and you do not need to spray again.

The sprays I mention have all been approved under organic growing codes adopted around the world and we accept their use for that reason. A final word on sprays – if you have a pest that you can't overcome by these gentle techniques and feel that you need to use something stronger, do so. But only spot spray. Don't blanket the area with the nasty spray and don't depend on its frequent use. Remember that you are growing grapes in an environment which has probably been abused and knocked out of balance with damaging sprays at least since the 1950s when these things came into vogue. Be prepared to take your time to set things right. It hurts having to compromise but it's better than going out of business before you get started.

After I had been on the local radio station talking about my attitude to sprays, several old timers stopped me in the streets of Mudgee and told me that I was doing what they used to do in the old days. They said I was right, that they had objected to the new sprays but had been howled down, mostly by their sons who came on to the properties with modern ideas.

In trellising, I am moving towards the idea of one post per vine, with a training wire at about 1.5 m. With a post to every vine, placed in the second year, you get help in training the vine and also provide resistance for a sensor, should you use one of the new mowing or cultivation devices with a sensing system.

Birds are a worry to farmers, mostly because farmers manage them badly. Birds are a part of life on the farm. You need to get to know them and know their feeding habits. We have lots of birds and have an annual census taken. The vast majority of the birds are insect eaters and they are in the vineyard because of our lack of harmful insecticide sprays. These resident birds tend to hunt off invading fruit eaters at harvest time, although it does not work so well in drought times when the invaders will run great risks. I believe the best long term strategy in countering bird attacks on your crop is to plant an alternative crop. This means identifying in advance the bird that will be the problem, then planting a diversionary crop. Figs are good. If your problem is silver-eye, plant a flowering tree, like the marri, which the silver-eye prefers. Scaring devices work – but only for a short period. They become bird educating devices. You have to keep changing them daily to be effective. I note that the West Australians have declared the starling a pest and have displayed posters in the buffer zone between their State and the east. I thoroughly approve of such tactics and wish that other States would have an open season on the starling, an introduced pest, and dispose of it the way the Chinese got rid of blowflies.

Another important element in control of losses to birds is in canopy design. Most birds don't like going into dense vines. They are always looking over their shoulders for enemies. The vines which get attacked are those which are far apart, sparse, and with grapes on show. Thus you need to encourage the development of large frames for your vines before you permit cropping. Prune to provide good leaf cover and concealment.

Another strategy is to provide water around the vineyard. Overseas tests showed that one third of the attacks on grapes were from birds seeking moisture.

Finally, you may simply have to go on patrol with a gun to educate them to feed elsewhere. Just remember that birds usually prefer food which is slightly acid, that is why their attacks on grapes, apricot and apples start before the fruit is ripe.

The well kept vineyard is a beautiful sight, the large green canopies of leaves providing shade over the grassy pasture beneath, a cool spot for a mid-summer rest in the middle of the day. The hum of friendly insects and the calls and songs of birds are an enchanting prelude to the feast of ripe grapes which awaits you in the autumn.

TOWARD AN AUSTRALIAN AQUACULTURE

By IAN "YABBY" CARSTAIRS

One reason for the great interest in aquatic animals as food is their potentially better food conversion efficiency than terrestrial animals. (Food conversion efficiency or conversion rate is the ratio between increase in body weight and amount of food eaten). Because their density is very close to that of water, aquatic animals waste comparitively little energy for their movements, and since they are cold blooded no energy is expended in thermo regulation.

Aquaculture also offers the possibility of optimising mans use of biological energy by shortening the food chain. Where biological energy very often passes two to four links before becoming available to man. Large amounts of energy are lost in its transfer from one trophic (nutritional) level to another.

The primary production of about 740,000 million tons of plankton algae per year yields a harvest of only 60 million tons of fish and other animals and plants from the seas. This is less than 0.01 percent of the basic production of phytoplankton,(small floating plants).

Since the marine production of food for human consumption involves about six times as much vegetable protein as the production of meat on land, it seems desirable to develop Permaculture activities which exploit short food chains.

In fish farming or in other aquaculture projects it is possible to shorten the food chains by choosing suitable species which feed on low trophic levels.

Mussels, polychaete worms and yabbies feeding on detritus (decaying organic matter) at the bottom of ponds are remarkably efficient energy converters. The same volume of water that supports an annual harvest of one kilo of Murray Cod produces 10kgs. of yabbies!

The aim of Permaculture is obviously to divert as much as possible of the attainable living matter of a pond ecosystem into channels that are useful to man, compared to those that prevail in a wholly natural food web.

Fertiliser needs, together with the advantages of full use of the water body and ecological resilience encourage multiple useage systems rather than intensive monocultures.

The following notes, covering freshwater, brackish and salt-water ecosystems and species are intended as a stimulus to regional research in aquaculture design for Australian conditions. The taking of measurements and the keeping of records by people in different parts of Australia will enable us to see patterns of movement and growth and ties between habitat and species which will suggest more productive integrations for the future.

TRADITIONAL AUSTRALIAN AQUACULTURE

The aboriginal tradition of fish farming goes back many thousands of years. In Victoria (at Lake Condah and Lake Toolondo)* remains of ancient fish farms have been excavated in the form of stone work. By building structures and transporting elvers, fish fry or yabbies to the naturally productive waters Aboriginals qualified as skilled fish farmers, using their highly developed observational skills to catch, trap and then go one step further than their efficient fisheries management into enhancing natural yields by fish farming.

Mary Gilmore's book *Old Days, Old Ways* describes the Aboriginal 'fish barrage' system. A log was precisely positioned to ensure that the flow of water over it was sufficient to prevent large fish swimming upstream, and not sufficient to wash young fish downstream. A lattice work of saplings under the logs allowed water but not fish to pass through.

Assuming that Murray Cod, Yellowbelly or some other large fish was the target, this separation of sizes was a basic step to reduce cannibalism and thus increase the yield.

These barrages were common before European invasion of the Murray River system. They were destroyed by Europeans wishing to take barges up and downstream. Later the lock system delivered the 'coup de grace' to Murray Cod spawning by the building of dams in the same sites.

Certainly the Aboriginal apprenticeship to fish farming involved years of careful obvservation; learning to feel the ecosystem while catching, handling and eating 'wild' fish. There can be no

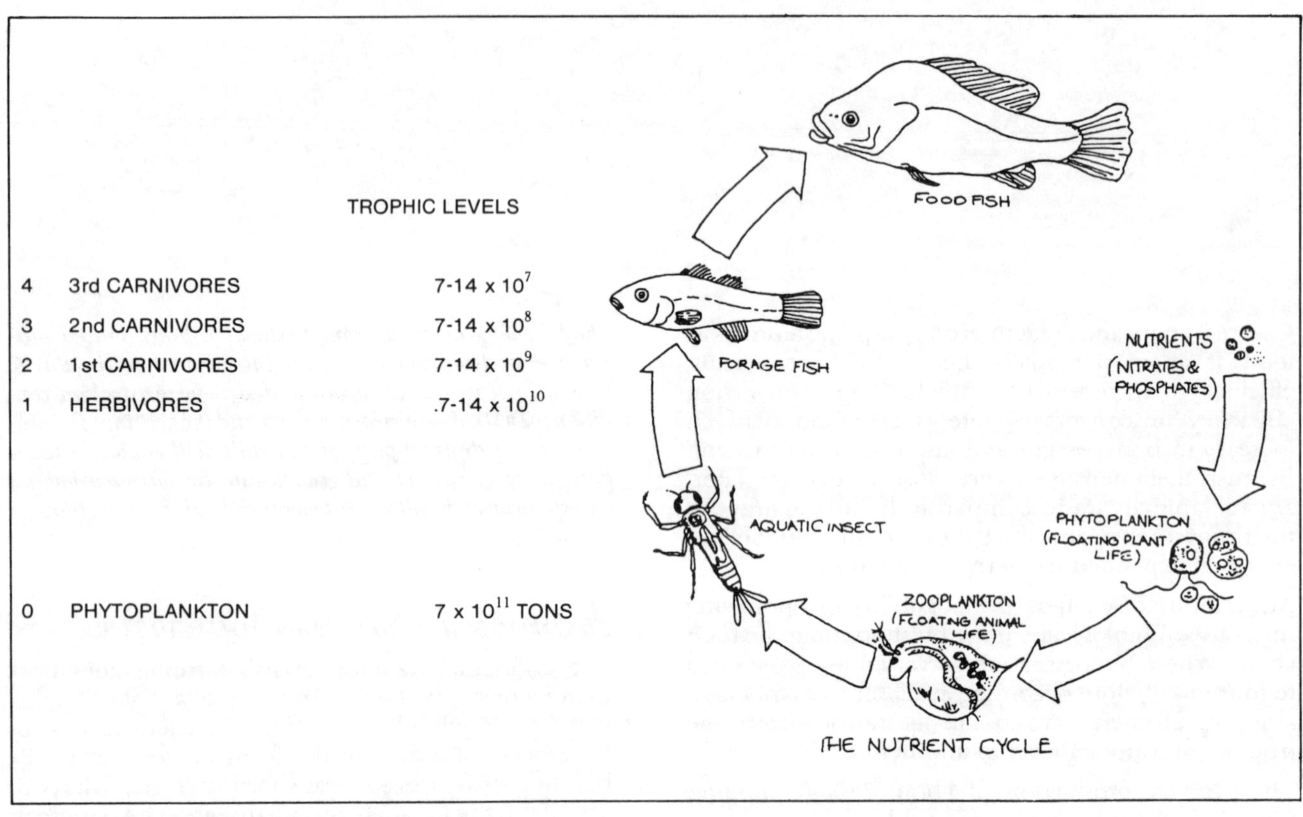

The production of organic material on different trophic levels in the food chains of the oceans. The ecological efficiency is estimated to be 10 to 20 percent in the first step (phytoplankton - herbivores). The ecological efficiencies in the subsequent steps are assumed to be 10 percent. (Source: M.B. Scheaffer, *Transactions American Fishery Society* 94 (1965): 123.

better recommendation for would-be fishfarmers than to take the same path before attempting to manipulate the environment of a species and improve on its yield from nature.

CONTEMPORARY POND CULTURE

THE NUTRIENT CYCLE

Whether on the land, in the sea, or in a fresh water body, the function of nutrients is identical. Only plant life, under the influence of the sun, can assimilate nutrients directly. On the land, green plants utilize the nutrients; in freshwater, two types of plant life carry out this function – firstly, the small floating forms of plant life which constitute part of the plankton (the plankton consists of small floating animals, zooplankton and small floating plants, phytoplankton) and, secondly, the fixed aquatic plants, the water weeds.

Whatever their source, the nutrients eventually become available to form the basis of the food cycle.

Animal manures from grazing stock, fertilisers from top dressed pastures and nutrients arising from the weathering of rocks may all find their way into ponds as a result of surface drainage within catchments.

The nutrients are assimilated by the floating algae (phytoplankton), the algae found on the surface of submerged rocks, and on aquatic plants, or by the aquatic plants themselves, and the food chain is commenced; nutrients have now become available to fish. Small fish species, and the adults of certain other species, may feed directly on the algae or on the aquatic plants, and here the cycle is short. Animal plankton (zooplankton) made up of forms such as the rotifers, the water flea *(Daphnia)*, Cyclops, feed on the phytoplankton; water snails (*Limnea*) damselfly nymphs, caddis fly larvae, feed directly on the algae found on the surface of submerged rocks and other under-water objects, or directly on the aquatic weeds. Small fish species (forage fish) feed on the animal plankton, on the shellfish and insects. When the algae, the animal plankton, the insects or the fish themselves die, bacteria commence the break-down processes and the cycle begins anew.

AQUATIC WEEDS

Weed beds are an important source of fish food; they offer protection to young fish; they offer spawning facilities to certain species of fish, and they assist in preventing erosion of the banks by wave action.

Weeds already growing in local waters may usually be safely transplanted to the fish ponds, but there are some aquatic pests which should not be introduced into farm dams unless appropriate management techniques are used. The major pests are red milfoil *(Myriophyllum verrucosum)*, water hyacinth *(Eichhornia crassipes)* and salvinia *(Salvinia auriculata)*. Cumbungi though providing good cover can spread rapidly and will readily fill in open water areas of constant depth.

Over-development of weeds may completely choke a pond. Weed-filled ponds are usually poor fish ponds and it is difficult, if not impossible, to fish in such a pond.

If the pond is not too large, the growth of aquatic weed can be kept in check by the use of a scythe.

As on land, tenacious plants can successfully be sheet mulched using plastic sheeting and gravel.

Quite a lot of aquatic plants will strike from cuttings of leafy branch tips and this may be carried out at any time during the growing season. Again shoots should be covered in clay, about 4 or 5 shoots (15-30 cm long) in each clay ball, the balls being dropped into the water.

Plant growth will normally follow the nutrients (other conditions being suitable), and insect, and other life, will follow in their turn.

Either as an adjunct to the weed beds, or in place of them, when weeds cannot be established satisfactorily, shelters of brushwood anchored down by stones or small logs, or merely heaps of stones, may be placed in the shallow water. These will shelter small fish, and also provide cover for aquatic insects, and yabbies on which the fish feed.

INSECTS

With most aquatic insects, the early or larval stage is passed in the water; this stage is the longest in the insect life history, and the most important from the viewpoint of a source of fish food. The adult form of many aquatic insects flies freely and may only return to the water to deposit its eggs; with others, the adult is aquatic, and is also capable of flight. Insects such as mayflies, caddis flies, damselflies, dragonflies, aquatic beetles, midges, and various water bugs such as water boatmen and back-swimmers will visit the pond and deposit their eggs.

This process may be assisted and accelerated by the addition of aquatic insects collected in other waters in the locality.

Trees should be planted around, and in the neighbourhood of the pond.

Flowering trees attract large numbers of flying insects in the warmer months and many of these will drop on the water, and they will appreciably augment the food

Table 1

Orders	Families	Common Names
Insects		
1. *Coleoptera*	*Chrysomelidae*	Leaf beetles.
	Dyliscidae	Water beetles.
	Dryopidae	Riffle beetles.
	Elaferidae	Click beetles.
	Gyrinidae	Whirligig beetles.
	Hydrophilidae	Aquatic and semi-aquatic beetles.
	Scarabacidae	Chafers (beetles).
2. *Diptera*	*Chironomidae*	Midges (larvae known as blood worms).
	Simuliidae	Sandflies.
3. *Ephemeroptera*		May-flies.
4. *Hemiptera*		Cicadas, planthoppers, &e.
	Corixadae	Water boatmen.
	Gerridae	Water striders.
	Jassidae	Leafhoppers.
	Notonectidae	Backswimmers.
5. *Hymenoptera*		Bees, ants, wasps & e.
6. *Isoptera*		Termites or white ants.
7. *Lepidoptera*		Butterflies and moths.
8. *Odonata*		
Sub-order *Anisoptera*		Dragon-flies.
Sub-order *Zygoptera*		Damsel-flies.
9. *Perlaria*		Stone-flies.
10. *Trichoptera*		Caddis-flies.
Other Groups		
11. *Crystacea*		Shrimps, yabbies, scuds, waterfleas, slaters, &e.
12. *Mollusca*		Shellfish, such as water snails, mussels, &e.

supply of the fish. Further, trees offer protection to water storages. They will shade the pond and assist in keeping the water cool.

WATER SNAILS, SHRIMPS AND FORAGE FISH

Other forms such as the various species of water snail, and shrimps will not establish themselves in a pond in the same way as has been described for aquatic insects. These will have to be collected in other waters with a shrimp net or simply by pulling up masses of water weed; the various forms, or at least many of them, will be among the weed.

The land-locked form of minnow *(Galaxias attenuatus)* and the pigmy perch *(nannoperca australis)* will live in still and near-still water. These small fish species serve as food for some of the species which can be raised in ponds (trout, English perch, blackfish etc.). The minnow normally will not complete its life cycle in still water. This species occurs in streams which run to the sea, and they move down to the mouth where they spawn in brackish water. The young fish move upstream to the freshwater, where they grow and develop.

If it is desired to stock a pond with forage fish, such as smelt and minnows, these can be collected by means of a shrimp net or a bait net.

PREPARING A POND

The actual depth or shape of a dam is important to consider. The longer the edge and the greater the shallow area there is, the more food is available. For this reason a gully dam is often excellent especially if it reaches to swamp areas, as the amount of edge and hence food is quite substantial.

Deep water is desirable for a number of reasons; the larger fish naturally move to the deeper water and deep water will ensure a region of lower temperature during the summer months.

Shallow water can carry a certain amount of weed growth which offers protection to the small fish, and which is also a source of large quantities of food. It is preferable if the shallow water can be provided by a sloping shelf in one section of the pond; shallow water should range from 1 foot to 3-4 feet in depth. The deep water should range from 5 to 12 feet or more.

It is desirable if the construction of the pond can be designed to allow for drainage if this should be necessary. This will permit the cleaning of the pond during the early winter months and also the complete harvesting of the fish population if this should be required at any time. (See 'Waterworks' page 107 of *Permaculture 2* by Bill Mollison for specific dam design.)

The nature of the pond bottom is extremely important in the biology of the body of water. Plankton, insects, and fish which die fall to the bottom of the pond; their bodies decompose and through bacterial action the nutrients again become available. On a 'good' bottom the decay is rapid; on a 'poor' bottom decay is slow. Further, on a 'good' bottom, the products of decomposition are absorbed and held by the mud, and are slowly given off over a long period.

Gravel, clay or sand bottoms can be improved by the application of stable manure or sewage sludge. Where vegetable debris has accumulated at the bottom of the pond without decomposing it may be necessary to add lime in large doses to bring about decomposition. The acidity (softness) or alkalinity (hardness) of a water is important. The term pH is used to express acidity.

DOSE OF CALCIUM CARBONATE (LIMESTONE)
REQUIRED BY MUDS OF DEFICIENT pH

pH of Mud	Calcium Carbonate Required in: Hundred-weights Per Acre
<4.00	30-60
4.0-4.5	24-48
4.5-5.0	18-36
5.0-5.5	15-24
5.5-6.0	8-15
6.0-6.5	7-8

The best pH for a garden is 5.5-6.0, but for ponds 6.5-9.0 is ideal.

The most useful form is ground limestone. Quicklime, which is sometimes recommended, consumes oxygen, flocculates clay, corrodes metal and kills fish and invertabrates!

STOCKING THE POND

Before a dam is stocked, it should be allowed at least three months to settle and allow the establishment of the food supply.

The number of fish placed in a dam for the first stocking should be related to surface area and not depth of water or total volume. It is the surface area that has the greatest relationship to the ability of the dam to produce natural food for the fish.

It is essential to avoid excessive stocking; this can result only in a high mortality or a stunted population. A given area of water will support only a certain weight of fish – large numbers of small fish or small numbers of large fish.

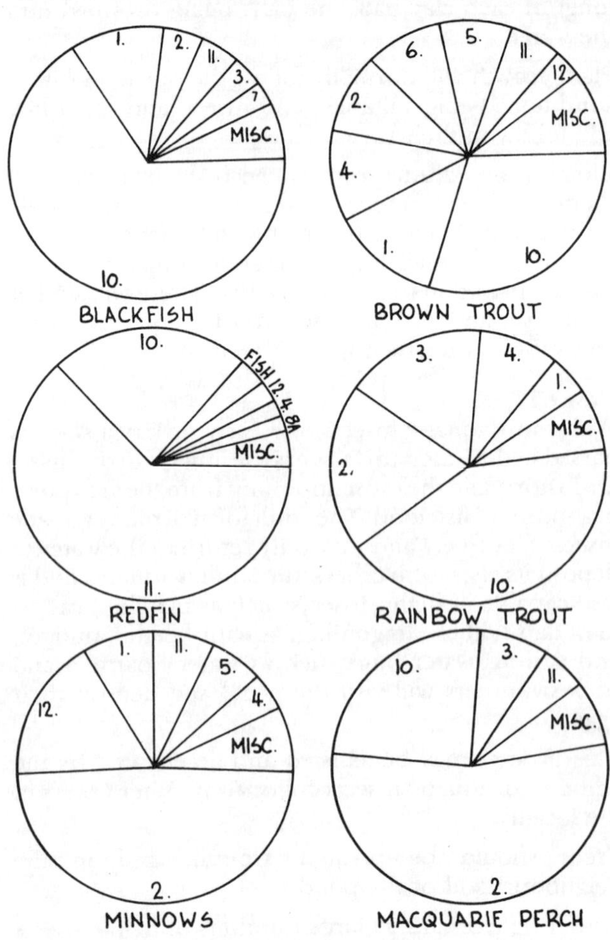

Dunbavin Butcher's analysis of the diet of 6 species of freshwater fish (see Table 1 for numbering key).

Care has to be taken as to the number and size of each species introduced; it is essential to maintain a proper balance.

State Fisheries Departments will advise on stocking rates and their advice is essential.

It is known that certain species interact very unfavourably with other species when they are associated together. Floods or other causes may transfer a species of fish from a pond into a water system where the introduction of this species is regarded as undesirable.

In Australia the European Carp is a case in point!

Applications to stock private waters with fish must be made to the appropriate person in each State.

When catching fish for transfer to a pond, the fish should be taken with a barbless hook. By this means, the fish will not receive any unnecessary handling, as they can be shaken off the hook into water in a suitable container, such as a polythene bag inside a plastic dustbin.

Handle, if at all, with wet hands. If the fish have to be transported over any considerable distance, particularly in hot weather, a bicycle pump should be carried and air pumped into the water at intervals, or whenever the fish show any signs of distress, such as gulping at the surface of the water. Before transferring the fish from the bin to the pond, the temperature of the water in the bin and that of the water in the pond should be approximately equalized. This can be done by immersing the bin in the pond and by tipping out small quantities of water from the bin and replacing it with pond water. A simple test such as dipping the hand into the water in the bin and into the pond is a sufficient guide as to temperature conditions.

INCREASING PRODUCTIVITY

Feeding types among fishes range from predatory gulpers to sifters of organic materials in mud, to zooplankton feeders and to herbivores that eat algae or even leafy plants. The aim is to select compatible species with different feeding patterns.

Among freshwater species the highest production of fish per acre is obtained with fish that feed to a large extent on phytoplankton. Species feeding largely on insects give the next highest production. Species that feed largely on other fish give the lowest production.

The fact that most of our Australian native fish fall into the second and third groups makes selection difficult.

However, it is known that fish learn to feed on almost anything and this provides scope for experimentation using cheap or near valueless crop residues.

Feeding should take place every day or at least two or three times per week at the same time during the day and at the same place. (Japanese Buddhist Monks signal feeding times by ringing a bell.) Start the fish off slowly with a small amount of food and gradually increase the amounts as the fish become used to feeding. Feed only what the fish eat rapidly as any uneaten food lies on the bottom, decomposes and can foul the water. Once supplementary feeding has begun it must be continued if the fish are to remain healthy. Do not feed if the water temperature is over 25 degrees C.

In 'controlled' production from ponds natural yields can be dramatically increased by the addition of organic manures which stimulate the growth of plankton, the basis of all food chains in water. In addition manures can be consumed directly by a variety of pond organisms including fish, altering the total dependence on light-limited plants which is a feature of natural ponds.

In Malaysia, pigsties are placed partially over the ponds so that pig wastes can be sloshed down into them. Water hyacinth (a noxious weed in Australia) is grown on part of the pond surface and utilised as pig feed and the fertile pond water is used to water market garden crops.

In Hungary ducks are stocked in ponds after fish have reached fingerling size and comparisons between fish ponds with and without ducks indicate far greater yields of fish from duck ponds. Sheds and runways for ducks are of course needed and ducks raised on fish ponds are healthier, leaner, cleaner and faster growing than ducks raised in more conventional systems. The fact that ducks eat organisms not ordinarily eaten by fish (e.g. aquatic weeds, frogs, etc.) leads to a more efficient utilisation of available feeds.

As well as providing nesting sites, islands increase the amount of shallow water and edge in the pond thereby increasing the food available.

Sewage utilisation, as in family or village privies over Asian fish ponds, is also worthy of consideration. Disease and parasite transmission, while a consideration, is often over-rated. Adequate pond management and careful cooking can overcome most of the potential problems.

PROBLEMS

The denser the stocking rate in Aquafarming the more difficult the management, largely because of the accumulation of waste substances, depletion of dissolved oxygen, and other problems of sanitation.

Poor water quality caused by accumulated metabolic products is a basis for many disease problems.

Care must be taken against overstocking with ducks or over fertilising lest excess organic matter bring about a depletion of available oxygen and anaerobic conditions in the bottom mud – nursery ground of many invertebrates that serve as important food components for the fish.

Fertilising should cease when visibility in the water is reduced to 20cm below the surface.

As wind and solar pumps become less expensive in relation to fossil fuels, pumping will increase even further the productivity advantages that Aquafarming has over terrestrial farming.

MANAGEMENT

The pond must be fished to remove the usable fish. The greater the number of pounds of fish removed, the greater the weight of fish the pond will produce. The food that was previously eaten by the fish that are caught is now available to those remaining. As already stated, a pond can only carry a definite weight of fish, 100 fish weighing 1lb. each, or 800 fish weighing 2 oz. each. If a pond is not fished, it will gradually deteriorate. Fertilizing and proper stocking must be followed by adequate fishing.

PROPOGATION

In the case of the rainbow trout it is most unlikely that any spawning will take place, and brown trout will spawn in a pond only under the most favourable conditions, viz., a gravel bed on a bank where the prevailing winds bring about almost continual and moderate wave action.

Trout are fairly easily "stripped". The eggs are a pale-yellow to a deep-orange in colour and are the size of a small green pea; the male product is white and is known as *milt.* Trout, both male and female, are stripped by flexing the body of the fish so that the lower surface is slightly arched outwards. The fingers of one hand are then lightly run along each side of the belly from the head and towards the vent, exerting a slight even pressure whilst doing so. If the eggs do not come away freely and easily, then the fish is not "ripe". The eggs are stripped into a clean dish containing a little water; the male fish is then stripped over the eggs and the contents of the dish are gently stirred with a large wet feather, and the eggs are fertilized.

In the case of trout, the fish may be removed from the pond and the above procedure carried out. A simple hatching box may be made from a shallow wooden case; knock the bottom out of the box and tack wire mesh of approximately eight meshes to the inch over the bottom. Weight the box so that it will float upright, and so that it has 4 to 6 inches depth of water in it. The fertilized eggs should be placed in this, and after a period of approximately three months (dependent on temperature) the young trout will hatch. A young fish on hatching (the alevin) carries a yolk sac from which it obtains its food for some time; in the case of trout this yolk sac is absorbed in five to six weeks. The young trout, now known as fry, should

be gently decanted out of the hatching box into shallow water in a weedy area. By this means it should be possible to maintain a population of trout in a pond.

Natural propagation of Murray Cod, Callop and Silver Perch has been accomplished at the Narrandera Inland Fisheries Research Institute, N.S.W. Although these fish are normally spring spawners the crucial stimulus was found to be not an annual rhythm, but a rise in water level (with a consequent plankton bloom to nourish the young) and temperature (different requirements for each species) "even Australian stocks of the European Perch, which in its native habitat, spawns in the early spring, regardless of water level, have been shown to respond positively to the addition of water to ponds. It appears that, under Australian conditions, flooding of dry ground has a more or less universal favourable effect on reproduction of fresh water fishes. Certainly manipulation of the water level should be attempted as a possible simple means of inducing or retarding spawning in any Australian fish species considered for culture" (Lake, 1967).

REFERENCES

* Records of the Victorian Archeological Survey, Number 7, June 1978 and Number 6, 1977.

Blainey G.
Triumph of The Nomads
For information on Aboriginal Fish Farming.

Gilmore M.
Old Days Old Ways
Angus & Robertson, Sydney (1934)

Lake J.S.
Rearing Experiments with 5 Species of Australian Fresh Water Fishes
Aust. J. Mar. Fresh Water Res. 1967, 18:137-152

#	Species	Freshwater	Brackish	Seawater	Cool/Temperate	Temperate	Sub-tropical	Tropical	Native	Hardy	Survive low °C	Survive high °C	Survive low O2	Turbidity tolerant	Still pond	Flowing water	Trophic level	Flesh quality	Poly cultural potential	Groupings with other species	#
1	Golden Perch	•			•	•	•	•	•	•		•	•	•			0/3	A	A	2	1
2	Silver Perch	•			•	•	•		•		•	•	•	•			0/2	B	B	1	2
3	Murray Cod	•			•	•	•	•			•	•	•	•			0/4	A/B	A		3
4	Freshwater Catfish	•			•	•	•	•	•	•	•	•	•	•			1/3	A	A		4
5	Macquarie Perch	•		•	•			•			•	•	•				1/3				5
6	Black Fish	•		•	•			•					•	•			1/4		B/C	NOT TROUT.	6
7	Bony Bream	•	•		•			•					•	•			0/4	C	C		7
8	Minnow	•	•		•		•	•			•	•	•	•			0/2		A	FISH FOOD SOURCE	8
9	Smelt	•	•		•		•	•			•	•	•	•			1/2		"	"	9
10	Pigmy Perch	•	•		•		•	•			•	•	•	•			1/2		"	"	10
11	Freshwater Crayfish	•			•		•	•					•	•			0/2	A	B/C		11
12	Yabbie	•	•		•	•	•		•	•	•	•	•	•			0/2	A	A		12
13	Marron	•			•	•		•	•		•	•	•	•			0/2	A	A		13
14	Giant Freshwater Prawn	•	•				•	•			•	•	•	•			0/2	A	A		14
15	Carp	•		•	•	•		•	•	•	•	•	•	•			0/2	B	A		15
16	Short-finned Eel	•	•	•	•	•		•	•	•	•	•	•	•			2/4	A	A		16
17	Tench	•	•		•	•			•	•	•	•	•	•			1/2	C	A		17
18	Rainbow + Brown Trout	•		•	•	•					•	•	•	•			1/4	A	A/B		18
19	Redfin	•		•	•	•			•		•	•	•	•			1/4	B	B/C		19
20	Barramundi	•				•	•	•				•	•	•			2/3	A	A/B		20
21	Giant Perch	•	•	•		•	•	•				•	•	•			1/4	A	A/B		21
22	Sleepy Cod	•					•	•				•	•	•			0/3		B/C		22
23	Milk Fish	•	•	•			•	•	•	•		•	•	•			0/1		A		23
24	Sea Mullet	•	•	•	•	•	•	•	•	•		•	•	•			0/1	B	A		24
25	Bass	•	•		•	•	•		•				•								25
26	Spotted Whiting			•			•	•	•				•				1/4	A	B		26
27	Greentail Prawn		•	•		•	•	•	•	•			•				1/2	A	A/B		27
28	Mud Crab			•			•	•	•				•				0/3	A/B	B/C		28
29	Sand Crab			•	•	•	•	•					•				0/3	A	A/B		29
30	Southern Rock Lobster			•	•	•		•					•				0/2	A	A/B		30
31	Sydney Rock Oyster		•	•	•		•			•			•				0	B	A		31
32	Pacific Oyster			•	•	•	•			•			•				0	B	A		32
33	Edible Mussel		•	•	•	•		•					•				0	B	A		33
34	Black Lip Abalone			•	•	•		•	•				•				1	A/B	A/B		34
35	Commercial Scallop			•	•	•		•					•				1	A/B	B		35
36	Pipi			•	•	•		•					•				1	B/C	C		36
37	Cockle			•	•	•		•	•				•				1	B/C	C		37
38	Squid			•	•	•	•	•	•				•	•			1/4	A/B	A/B		38
39	Spirulina	•	•		•	•		•		•			•				0	A	A/B		39
40	Cow Hair			•	•	•	•	•	•				•				0	A			40
41	Sea Grapes			•		•	•	•					•				0	B			41
42	Ogo		•	•	•		•						•				0	B			42
43	Kelp			•	•	•		•					•				0	B/C			43
44	Fairies Butter	•	•	•			•	•			•		•				0	B/C			44
45	Sea Lettuce		•	•	•	•	•	•	•				•				0	B/C			45

RULES FOR THE KEEPING OF A SILK-WORM FARM

CHINESE MISCELLANY.

1. Plant a hedge all round the farm. During the winter plant twigs of willow and ash mixed and when they have begun to grow plait them closely together, binding them with fibre of the coir-palm; outside the hedge plant sour dates, small oranges, thorns and briars. The more prickly the plants are the better.

2. Set up a sign with the words, 'Establishment for the encouragement of silk cultivation,' and ask the Mandarin for a notice forbidding people to tread down plants.

3. If you can lead the river water to irrigate your field, you will save the expense of digging wells; while at the same time you can keep fish and breed ducks in your canal or cultivate the water lily, water chestnuts, etc.

4. In the winter months you must nourish the roots of your plants, for which purpose make use of rotten fish preserved in jars, or the juices of decomposed vegetables.

5. In the fourth month plant your mulberries; in the following year, besides gathering your mulberry leaves, should there be any vacant ground, you can plant it with arbutus, apricots or pears.

6. Beneath the mulberries you can plant onions, leeks and melons, or potatoes and yams. In order to prevent the locusts from eating mulberry leaves, you may plant more yams.

7. Withered mulberry leaves, the parasites of the mulberry, together with white and diseased silk-worms, may all be used as medicine; but the snails found on the mulberry must be carefully caught, lest they injure the leaves.

8. In the establishment there is need of ladders, tables and sieves, all of which are made of bamboo; hence it is necessary to cultivate that plant.

9. The leaves of the old mulberries are good for fattening goats, and goat dung is good for feeding fish. Hence goats may be kept with advantage.

NOTE:
These directions have been carefully followed by the establishment issuing them, which in two years was already prosperous.

Edwards, E.D.
The Dragon Book
(1943 reprint – 1st Pub. 1938)
William Hodge & Co. Ltd London

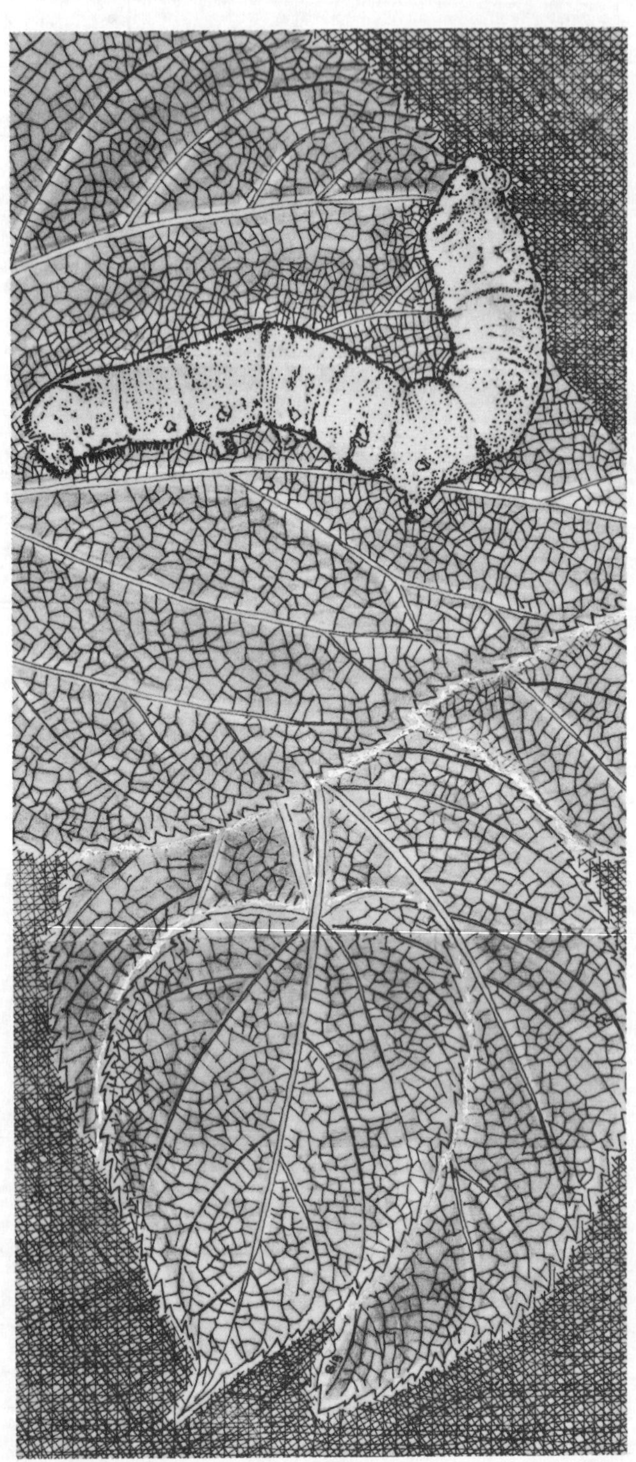

PIGS AND TREES

By RICK BRUCE

We live on a seventy acre farm at Darradup near Nannup (33.57 S, 115.42 E), Western Australia, and have been applying permaculture for about four years. Although when we first started, our ideas weren't so evolved and the permaculture movement has provided much inspiration since then.

Our farm was used for grazing beef cattle five years ago with not very well established pastures and since then, with no grazing or very little, the bracken, wattles (prickly moses), flooded river gums, marri and jarrah have been reafforesting very well. The bracken is very thick and high in parts, pioneering and conditioning the soil for the return of the forest.

Since March this year, we have discovered pigs and their potential. The pigs have been used to dig out patches of mainly bracken, and we have realized how well they do it and more. We started off with two pigs at the weaner stage in a dismantleable weld-mesh cage approx 8' x 10' which is moved around as required after each piece of earth has been 'rotary hoed'. After we have had the weaners for three or four weeks they are moved out of the mobile cage into a suitable sized area surrounded by a solar electric fence which works very well. The electric fence is made up of a nylon cord inter-wound with a strand of stainless steel which is carried and insulated on steel star pickets which are easy to move around. The pigs soon learn after a few touches and won't go over or through the fence. The solar unit stores the sunlight energy in a solid gel battery and has not yet lost its charge even after a week or more of continuous cloudy weather. If for some reason the fence isn't working the pigs will still not go through. So the pigs are preparing the earth by digging and fertilizing as they go, for no cost at all. This beats a tractor, rotary hoe or any other machine, except that the pigs need constant care and feeding.

We kept our first two pigs for four months or so and sold them to people who wished to buy them for meat. The money from the sale of these pigs was then used to buy two more pigs and will supply their food and perhaps a little more which can be used for supplying more resources such as trees and seeds. The amount of money made depends on the amount of food you can provide from your own farm or environs.

Pig control – the electric fence.

The area the pigs are working becomes attractive to the chooks who follow, often recycling the pigs faeces, foraging and manuring too. The earth is made softer and is growing a healthy growth of grasses and clover which can be used for mulching the surrounding trees in the spring. As the pigs grow and dig their way happily we can follow happily too planting trees, vegies, herbs and so on. The pigs work efficiently on the bracken growing in sandy loam and loam. They dig up the healthy bracken roots to eat and will use the bracken fern to make their own bed, preferably under a small shelter of shrubs or trees to shelter from the rain or wind. Trees of value should be protected from the pigs. The dead bracken then makes an excellent mulch rich in potassium and very fibrous taking a long while to go 'into' the soil. Their efficiency on gravel or hard clay would probably depend on the season the ground is worked, as they dig better on moist or rain soaked ground. Any bracken regrowth can be hand dug or left for a second dig by the pigs at a later date. We are now using our second two pigs and plan to eventually use them to prepare the ground for permaculture over the whole farm. Once you know pigs they could be used to breed your own supply of endless diggers and fertilizers. We are going into it slowly, learning the needs of pigs to make sure they work efficiently. They like room to move and graze, to forage, a protein supply, garden weeds and vegies, and being fed two meals a day of a grain/pollard/protein ration. Dead kangaroos and emus etc. hit by cars make good tucker, cooked of course.

That is our introduction to pigs, a beginning and we have a lot to learn.

We have planted many trees with success. Planting them out usually in the autumn/winter/spring. They receive no, or very little water in summer except those near a tap and those showing signs of dehydration. With this method we have had very few failures, the most important factor being care when planting and some mulch and some compost dug in around the roots if available. No irrigation for the permaculture forest is planned except for a small orchard area and citrus trees. The trees find their own water supply and will produce satisfactorily with adequate food supply. Some situations where they won't tend to dry out in summer. Run-off from baths, sinks etc. are very useful for trees and gardens.

To produce a garden of Eden for all is our immediate aim and the challenge as Fukouka says in *The One-Straw Revolution. The ultimate goal of farming is not the growing of crops but the cultivation and perfection of human beings.*

Steve Walden of Palmwoods, in S.E. Queensland uses a very similar method.

Land cleared by pigs, it used to look like the area in the background.

Pigs and trees at Steve's property. Remarkable changes in ground cover. Erosion control on slopes is essential. Note ground cover sown on treated area – left.

FOREST REGENERATION

By MAX O. LINDEGGER

Many properties on the Sunshine Coast have been overcleared. The size of rural blocks often makes conventional re-afforestation too expensive. With a tree density of 600 - 1000 trees per ha the cost of soil preparation, trees, some outside labour and other associated expenses put the cost per ha at about the $2,000 mark - and out of reach of many landowners. Cooperation, sharing of labour and equipment can reduce the cost. Tax incentives are available for shelter belts and tree planting in association with erosion control. A pool of willing voluntary labour is at times ready to take on the task.

Still, the job ahead is huge and any method which could make the work easier is worth looking into. The method described below has been adapted from the experience people had in South Australia and Victoria.

Natural generation has always been able to keep pace with natural loss. What changed the situation was the introduction of grazing animals. Cattle and sheep do not allow any new growth - What's not eaten is trampled. *Grazing animals are the number 1 enemy of any young plant and weed competition can be enemy number 2.*

In South Australia it is recommended to spray areas set aside for natural regeneration with a herbicide. Another method is the use of fire, and a third clean cultivation. None of the above methods can be recommended in our subtropical climate. The gain of young trees would be more than offset (negatively) by the loss of soil due to erosion which would be invited by clear paddocks.

Since the native seedlings must compete with weeds and grasses for moisture and nutrients, some help must be given. It should be understood that summers with good and well distributed rain are likely to be most successful. Dry summers are bound to be disappointing.

HOW TO DO IT:

The area set aside for natural re-generation needs to have some remaining trees of desirable species. These trees are known as seed trees and have usually been left as shade trees. The area has to be securely fenced. The fence should be placed out from the base of the

Young eucalypt regeneration.

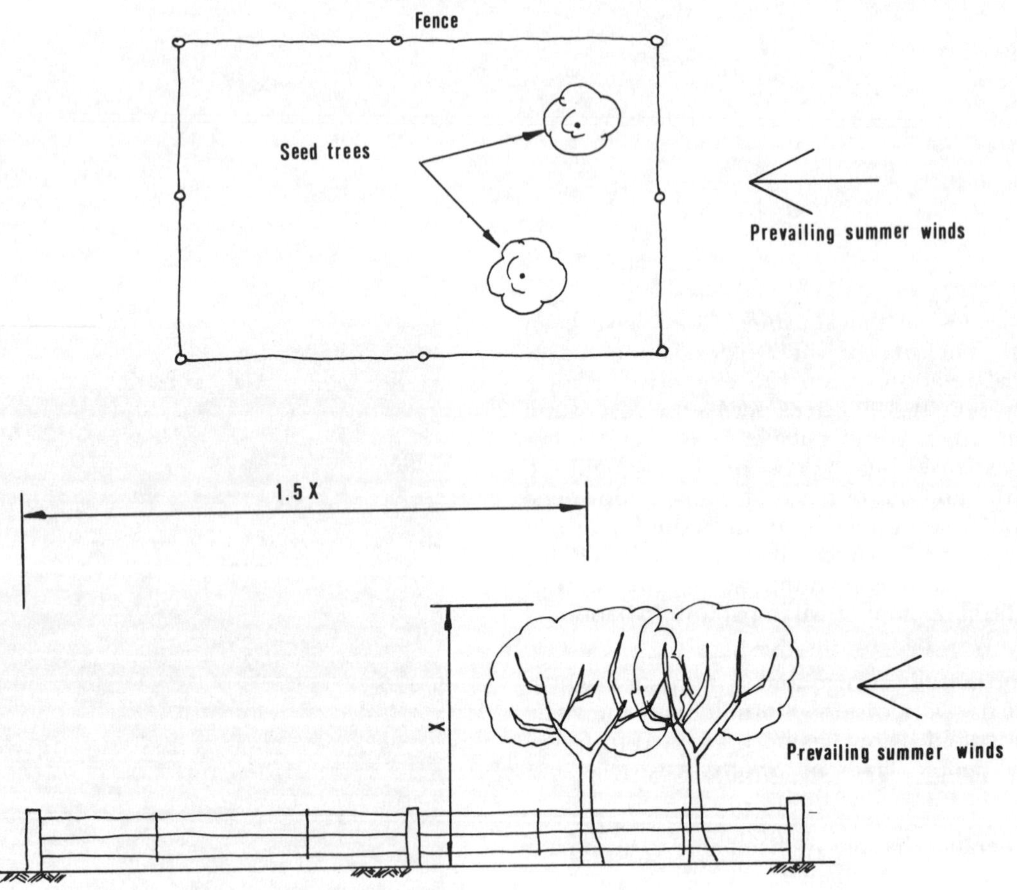

trees at least one and a half times the height of the seed trees in the direction of prevailing summer winds. This will ensure that a large number of seeds will fall within the enclosure. Most eucalypts drop much of their seed in late summer and the direction of the winds at this time should be used for determining the position of the fence.

Before fences are put up, it would be of value to aerate and condition the area with a chisel plough or a Wallace soil reconditioning unit. Following seed fall the grasses and weeds should be slashed as closely to the ground as possible. Weeding around emerging seedlings will increase survival. Hares should be controlled and if possible eliminated from the area to be re-forested. Maintenance of the fence is a must.

Once the new plants are established and are no longer prone to excessive damage from grazing, the fence can be removed and the treatment repeated around another group of trees. Depending on soil and environmental factors, this should be a period of a minimum of about 2 years and a maximum of about 5 years in most cases.

ADVANTAGES OF THIS METHOD:

With this method seed is supplied by local trees which are well adapted to local conditions and thus give a better survival rate. The method is well suited to rough country or small pieces of land too far from the home to be effectively controlled. Above all, this method is very cost effective - an important factor on most farms.

The hills and gullies in the Sunshine Coast hinterland have changed dramatically in the last 100 years. The replacement of the cover will take some time and persistence, but it will be well worthwhile.

COMMENT

While the above article was written for a subtropical climate, experience has shown the methods to be applicable in any area where rainfall is sufficient and heavy downpours have to be expected.

FOOD FORESTS LESSON ONE, UNDERSTANDING WHAT IS THERE

AUSTRALIAN NATIVE FRUITS.

By PETER HARDWICK.

"Changes" as Fukuoka notes, "must first come at the basic philosophical level". The changes I seek are very much a matter of philosophy, a search for the right question. Of the two questions — what can I demand this land to do? or what does this land have to give me? The first leads to a forcible rape of land by machinery, and the second to a sustained ecology supported by the intelligent control of man...It is a philosophy of working with, rather than against nature; of protracted and thoughtful observation; of looking at plants and animals in all their functions, rather than treating any area as a single product system...To understand, to use what is there.

Bill Mollison, Permaculture 2, page 2

Permanent cultures are site-specific. A unique blend of people, place and plants. Peter Hardwick is a horticulturalist and edible native plant enthusiast who runs courses on the north coast of New South Wales, Australia. When Peter considers the food forest he starts with what is there.

Our North Coast region is naturally endowed with a large number of native edible plant species. Other than the macadamia, their potential as developed food plants generally has been unrecognised.

Even without the selection of cultivated varieties, many local plants produce foods with exceptional flavour. One could predict that the next generation of food plants will include more of our local trees.

The region's moist sub-tropical climate, geography and range of soil types have produced a variety of natural habitats, each containing its own unique flora adapted to specific environmental conditions.

Each continent has provided its quota of developed food plants. Europe, Asia, the Americas and the Middle East being the major centres from where most of our present food plants originated.

Though we have at least 20,000 plant species in Australia, we have managed to select and develop only two species for commercial food production: The smooth and the rough macadamia (*Macadamia integrifolia* and *M. tetraphylla* respectively).

The tribal Aboriginal people survived very well on the local bush tucker for at least 40,000 years. Their diet included a healthy proportion of food from local plants. They had excellent knowledge of where the most palatable varieties grew, when they would be in season and how to prepare them for eating. Most of our knowledge of edible native plants comes from the Aboriginal people.

The reason for there not being a range of natives included in our modern menu must in part go back to the unaccepting palates of our colonial ancestors. They saw the edible native plants as a poor substitute for the introduced, highly developed food plants they had been used to.

Only at times of food shortage were native fruits eaten – usually very reluctantly. This most probably reinforced the image of the native food as being inferior.

LITTLE RESEARCH

Not everyone shunned the potential of edible natives. J.H. Maiden, an eminent Australian botanist of the late 19th century, included a section on human foods in his still popular text, *The Useful Native Plants of Australia.*

But since then very little research has been done on selecting suitable varieties of edible native plant species, macadamias being the exception.

To be fair to our colonial ancestors and scientists of more recent times, there is a major problem with researching edible native plants: The basic lack of knowledge on what is edible and what is not.

Undoubtedly, many intrepid native food connoisseurs end up with stomach ache or worse from eating unpalatable or poisonous plants. Correct identification and extensive knowledge of the plant is a must before any part of it can be eaten. The fundamental rule is: "If unsure, don't try it!"

I always remember the ethic of, "Never sample a native plant where it results in the destruction of the plant," such as edible tubers or yams. Whenever I harvest native fruits in the bush I always leave plenty of fruit so the plants can reproduce and wildlife can still get a feed.

Wanton destruction of natural vegetation is bringing many species to the verge of extinction, and the relentless spread of urban sprawl and conventional agriculture displaces and destroys many others.

Reliance on a small number of plants carries great risk, as monocultures are extremely vulnerable to catastrophic failure, brought about by disease or variations in climate.

To help feed, clothe and house a rapidly increasing world population, it is timely to consider neglected or little known indigenous plant species.

To this end, the number of botanic gardens, field stations, and habitat reserves containing natural vegetation types must be increased. Careful preservation and thorough cataloguing are particularly important for little-known plants.

Only in this way will the genetic diversity and healthy stock needed for new food plants be assured.
Potential breeding stocks, clones and cultivars will otherwise become extinct.

The variety and potential of the indigenous edible is there, but greater general awareness of their food value has happened only recently. A few popularised publications such as *Wild Food in Australia* by A.B. and J.W. Cribb have helped.

As is well recognised, research is needed urgently to increase the yield of these food plants. I can't help but wonder how many taste sensations we are denying ourselves by not utilising our edible native plants.

DAVIDSONIA SPECIES

The two local species of Davidsonia are rainforest trees with excellent potential for cultivation as food producers. One, Davidson's plum grows to 12 metres and has large ornamental leaves.

The Smooth Davidsonia is an unnamed, rare tree with smaller, hairless leaves. It is restricted to a few locations on the North Coast of N.S.W. It is known to grow only from cuttings, for the seeds are mysteriously sterile.

These Davidsonias bear prolific quantities of plum-size purple fruit with crimson edible flesh. They can be used for jams, pies, jellies and juices and are said to make excellent wine.

LILLYPILLIES

Whenever I go bush I eargerly seek the tangy, sometimes, aromatic fruit of the lillypillies (*Acmena* and *Syzgium* species), a welcome change from cultivated foods.
These plants belong to the same botanical family as the better-known eucalypts. The common Lillypilly (*Acmena Smithii*) is a tree which grows in coastal forests from Gippsland (Vic.) to Cape York (Qld.).
As with many other native plants, there is great variation in the quality of its fruit.

Davidsonia. (Photo, author).

The Brush Cherry *(Syzgium paniculatum)* a native of the Big Scrub, usually is said to have the highest quality fruit of its group, with a distinctive flavour all of its own.

The Blue Lillypilly *(S. coolminianum)* and the Riberry *(S. leuhmannii)* also produce exceptional fruit. Fruits from the Lillypillies can be eaten raw, in jams and jellies. Their close relatives in Indonesia and South America are some of the staple fruits for the people of these countries.

The Bauple Nut *(Hicksbeachia pinnatifolia)* is closely related to the macadamia, whose foliage it resembles, but with larger leaves and more striking appearance. It too has a fine-quality nut. The tree's ornamental value and small size (6m high) make it an ideal tree for the backyard.

Macadamia.

128

The Native Tamarind *(Diploglottis australis)* is a distinctive rainforest tree, its canopy crowned with large coarse leaves. Its fruit is very sour-tasting when eaten raw, but excellent as a fruit drink when diluted with water and sweetened to taste.

Native Tamarind, branch free to the high crown.

The Lacebark Kurrajong *(Brachychiton discolor)* and the Flame tree *(B. acerifolium)* are two local trees commonly grown as ornamentals. Their seeds are edible and can be eaten raw or roasted and ground as a substitute for coffee.

The Plum Pine *(Podocarpus elatus)* has been described as having among the best of the indigenous fruits. The tree is elegant and renowned for its fine-quality timber. Its fruit can be eaten raw, and provides a tasty difference in a fruit salad. The fruit's mucilaginous nature also lends it well to jellies.

Like the Plum Pine, the Black Apple fruit *(Planchonella australis)* needs to be eaten when thoroughly ripe. The fruit are large, almost black, plum-like up to 5cm long, sometimes found in considerable quantities on the rainforest floor. When soft and ripe I find their flavour like a cross between guava and custard apple. With a little variety selection, this could easily become an excellent dessert fruit.

The huge cones of the Bunya Pine *(Araucaria bidwilli)* contain large seeds about 5 cm long. These are said to be very tasty boiled or roasted, and were considered a delicacy by Aborigines.

Though rainforests contain the greatest range of vegetation, other habitats in this region also include edible plants in their pantries. Wetlands, woodlands, sand dunes and heath, all have their delicacies.

For the layperson who wishes to identify the rainforest flora there is a rainforest identification kit by J.B. Williams, published and distributed by the University of New England.

FINGER LIME

Few would realise that our region's rainforests contain an important rare citrus-type plant, the Finger Lime *(Microcitrus australasica)* which is being used to breed new and better citrus root stocks. The Finger Lime is not quite the store-bought orange or lemon, but rather a delicate cylindrical fruit to 6 cm growing on a thorny shrub or small tree. The first settlers to the region used the fruit extensively by making a fruit juice or sucking it to quench their thirst when they were in the bush. It is said to have a distinctive flavour with an oily after-taste.

The introduction of the domestic citrus resulted in the phasing-out of the usage of the Finger Lime. Clearing of the rainforests, introduction of citrus pest and diseases and over-grazing by cattle resulted in the plant becoming rare.

More recently, research has focused on the Finger Lime's resistance to the diseases, citrus root-rot fungus (Citrophthora) and nematode. Dr. Don Hutchinson, a research geneticist with the US Department of Agriculture, has visited the region on several occasions to collect specimens of the Finger Lime. He has been working on developing a viable citrus rootstock which will also have the disease resistance. Initial investigations failed to find suitable specimens, though eventually, with the help of local people, 50 or more specimens were located within the region's rainforest remnants. The prize specimen was a Mullumbimby strain which proved to be 100 per cent resistant to citrophthora. The pink-fleshed variety of the Finger Lime *(M. australasica* var. sanguinea) was considered a significant find as it may prove of value in crossbreeding with existing 'blood oranges'.

The history of the Finger Lime is an example of the importance of many of our rare and edible indigenous flora.

The Peanut Tree *(Sterculia quadrifida)* is a rainforest tree that you would be unlikely to find growing readily in the local bush, as it is rare in this region. The 'seeds' have a nutty taste similar to a peanut. The Peanut Tree loses its leaves each year, and has woody leathery pods which split at maturity to reveal a beautiful, orange-red inner surface with a row of up to five satin-black seeds. This tree is more common along the Queensland coast. As the Richmond Valley is the most southerly extent of the species, preservation of local strains is important as they may have greater cold resistance than tropical strains.

The Durobby *(Syzigium moorei)* was formerly common throughout the Big Scrub area and its edible fruit was favoured by Aborigines. The usual arrangement of red fruit following white flowers is reversed in this handsome tree. Showy red flowers are borne on short stalks from the old branches rather than the young growth as most other species, and the large, rounded fruits, about 5 cm across, are cream-coloured.

Durobby. (Photo, author).

There are several other edible native species that are rare or endangered. But the development of a food plant also requires more than just the preservation of the species. It requires the preservation of a broad range of genetic types within each species.

Some of the edible native species which are rare or endangered are:

The Small Bolwarra *(Eupomatia bennettii),* has a small guava-tasting fruit likened to a rose-hip to look at;

the Small-leaved Tamarind *(Diploglottis campbelli);* and the Narrow-leaved Orange Thorn *(Citriobatus lancifolius)* are very rare rainforest species with edible fruit. The largest specimen of Narrow-leaved Orange Thorn (17 metres high) is growing in the Rotary Park rainforest reserve, Lismore.

The Cobiotics Foundation has a list of local edible native species which are considered to have good potential for food production systems. For a copy of the list send a Self addressed and stamped envelope to the Cobiotics Foundation, P.O. Box 26, Lismore 2480.

Food value is only one of the practical applications of indigenous flora. Native vegetation has many useful properties, including medicine, timber, animal forage, dyes, tanning, fuel, ornamental and many other uses.

FLOWERS FOOD FOR THOUGHT

By BRIAN STOCKWELL

Invariably, after I have given a talk or shown a group around the garden, a lady will come up to me and say, "I'm glad you said that about flowers, I've been trying to persuade my husband into letting me put some in the vegie garden". She would be referring to my pet phrase "Food may feed the stomach but flowers feed the heart".

But is that all they do?

Flowers and flowering trees and shrubs are an integral part of any garden or agricultural system. Not only do they add a splash of colour that brightens the gardener's spirits, they also may be a nutritious food source, a pest deterrent or growth accelerator, a forage for bees, the basis of a refreshing beverage, or a powerful medicine.

Some may think it odd to consider eating flowers but in Elizabethan times it was not uncommon to talk of stewed primroses or gillyflower cordial. The rose was classified, until recent hybridization for looks alone, as a culinary vegetable. Let us look at some flowers that may be used in cooking or eaten au naturel.

Being the year of the tree, we'll start with the humble hibiscus tree. In fact not only the flowers, but the leaves and most parts of the tree are edible. Animals eagerly seek its leaves for fodder. The flower has a delightful taste and is an abundant source of highly nutritious pollen.

Carnations – For centuries carnations, often called pinks, were prized for their clover-like flavour and fragrance.

Chrysanthemums – The Chinese favour this flower with a ceremonial dish served only on special occasions.

Dandelions – Children love to find dandelions, if only to make a wish as they blow off the fluffy heads. Every part of the dandelion is edible. It's inner leaves are delectable in salads. The blossoms make dandelion wine and the root roasted and ground is a good coffee substitute.

Marigolds – Marigolds are popular for their ability to repel nematodes in the soil. However, marigold petals

Nasturtiums.

on aluminium foil with none overlapping and put in the oven for two hours at 100C or until they are dry enough to crumble, store in airtight containers for later use (finely crumbled) to colour noodles, rice, soups, etc.

Nasturtiums – As a member of the cress family it has similar taste to water cress although much more delicate. The flowers (and leaves) are delicious in all salads, alone or with greens. Nasturtiums are also a good companion for brassicas, to help balance the effect of selective breeding which has produced the large terminal bud we know as a cabbage. The scent also deters aphids from brocolli.

Roses – are both beautiful and rich in vitamin C (more so than orange juice). The fragrance will enhance jellies and jams. The rose needed is the old fashioned variety which is more delicate. Modern strains have less fragrance, are stronger tasting and have tougher petals. The thorns of the rose and its bushy nature make it a good border plant.

Sesbania grandiflora.

Mallow.

Violets – A member of the herb family, the violet is not only appealing to the eye but also to the palate. Both blossoms and leaves can be eaten.

Native Trees – Although a number of older flowers of cultivation have been put to culinary use for a long time, both in herbal remedies and recipes, there has been little similar tradition connected with our native flowers, but the possibilities are here!

Acacia - Blossoms can be made into fritters which contain a copious amount of pollen.

Grevillia – The delightful blooms of the grevillia can make a delicious tea full of nectar.

Tea Tree – Aborigines were observed by early pioneers to steep tea-tree blossoms in water to produce a subtly flavoured beverage.

Native Hibiscus – Growing prolificaly on the Sunshine Coast, it, like its tropical cousins, has edible flowers and leaves. It has the advantage of being a large tree that will grow in water-logged conditions.

Flowering species may protect vegetables and fruits from pests and diseases.

Marigold – may protect beans from nematodes, and the pea and bean weevil.

Zinneas – protect cucumbers, melons, squash from most pests.

Tagetes – Stinking Roger is by far the most useful in protection of other plants.

Other flowers have a beneficial effect on plants when grown in association.

Calendula – will attract parasitic wasps away from squash.

Dandelions – stimulate fruits or flowers to bloom more quickly.

Cornflowers – were so named because of the beneficial effect on corn crops.

Morning Glory – Although dreaded by some, it enhances root vigour in corn.

What a delight a multi-coloured countryside would be if farmers and gardeners could understand the inter-relationships of nature. So next time you have a "feeling", a "hunch", or an "inkling" to plant flowers amongst your "productive" vegie garden or orchard or vice versa – follow that inspiration. Nasturtiums will grow up tomatoes and protect them from insects, strawberries will mingle with lettuce, while garlic grows between rose trees keeping pests away. If you mix the pretentious formal garden with the vegetables, they express their joy by flourishing together.

The silent fame of flowers flows freely in realms yet to be explored and understood by the vast majority of gardeners and farmers. The field lies waiting for us to forget our forgone conclusions and to begin to find the favours and flavours of flowers.

REFERENCES

Another Way of Living – Jacques Massacrier
Esther Deans Garden Cook Book – Esther Deans
Flower Fairies of the Seasons – Caly Barker
The Good Food Growing Guide – John Bon
Health Secrets of Plants & Herbs – Maurice Messegue
Language of Flowers – Kate Greenway
Wild Food in Australia – A.B. & J.W. Cribb
Companion Planting – Helen Philbrech & R. Gregg.

SALVINIA – AN ORGANIC'S PARADISE

RAY PAYNTER

Aquatic weeds are causing great concern world-wide because of their ability to multiply and completely block water-ways and reservoirs, sometimes to the extent of doubling their area in 6-7 days as does Water Hyacinth. They have always existed, but have exploded with man's interference with nature where pollution in the form of run-off fertiliser has enriched our waterways. Aquatic weeds reflect the composition of the water, in that where the water is rich in run-off, weeds grow abundantly and contain many rich elements preciously needed in our soils especially those of our monoculturists.

For those interested in more technical information the analysis as follows was prepared by the Department of Primary Industry, Nambour, Queensland.

Element	Symbol	Value
Nitrogen	(N)	1.19%
Phosphorous	(P)	.174%
Potash	(K)	.200%
Calcium	(Ca)	.760%
Magnesium	(Mg)	.493%
Iron	(Fe)	10,000 ppm v.v.High
Manganese	(Mn)	2,820 ppm v.v.High
Zinc	(Zn)	59.4 ppm
Copper	(Cu)	12.8 ppm
Chlorine	(Cl)	.001%
Sodium	(Na)	201 ppm
Boron	(B)	11 ppm

pH value 6.5-7
ppm (parts per million)

It is nature's way of balancing and of recycling lost elements and all that man can think of is to spray it and get rid of it! We are realising the prospect of using weeds such as Salvinia and Water Hyacinth, which constitute a free crop of great potential value needing no tillage, fertiliser, seed or cultivation, only reaping.

Their uses are numerous and include animal feed, human food, soil additives, fuel production (methane) and waste water teatment. Our immediate attention is drawn to using these weeds as soil additives. In its dried form they have greater nitrogenous content than pure chicken manure and possess numerous minerals and trace elements. They are primarily, extremely efficient as a mulch; conserving water loss, preventing weed growth, increasing earthworm activity, helping prevent erosion, keeping soil temperature more even. They even have good visual appeal with no smell from decaying weed.

Overseas, in the U.S.A. and many underdeveloped countries, experiments with extraction of aquatic weeds are being carried out successfully using varying methods according to the size of the problem. Most

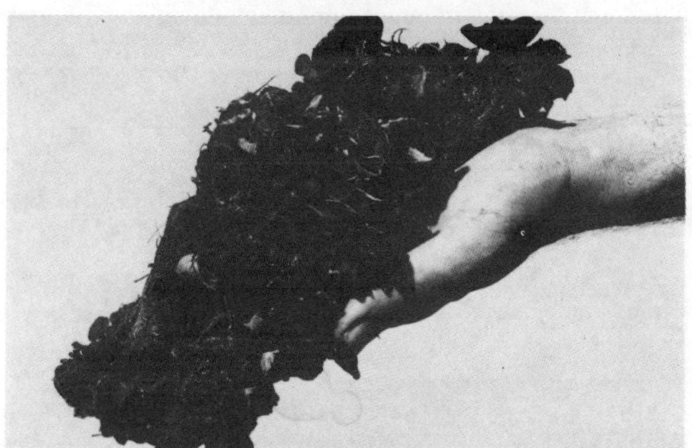

however, operating on an elevator system either from shore or from a pontoon-boat like structure. The water content is however the major drawback, being in the region of 90-95%. This has been tackled by chopping the weed first, reducing volume to approx. 25% then pressing, where 50% moisture can be extracted very easily. The compressed weed can then be handled economically.

Water Hyacinth grown on sewerage effluent has multiplied from 2 plants to 30 in 3 weeks and 1200 plants in 4 months and has been harvested at a rate of 800kg dry weight per hectare per day.

In methane production, 1 hectare has produced 10,000 cubic metres. Biogas production removes carbon by fermentation but little else, and leaves excellent sludge useful as a fertiliser.

Some aquatic weeds have the same amount of protein as soybean and if harvested regularly, crude protein produced in 1 hectare equals a 60 hectare crop of soybeans.

It seems for the next twelve months anyway we have a reprieve from councils eradication programme which more than likely would involve spraying your water supply with a mixture of kerosene, detergent and a herbicide F101, the combination of which, would sink the weed and allow it to rot under the water causing increased pollution and the production of 'swamp gas'.

COMMENT

In many areas Salvinia is not a problem anymore due to the introduction of a biological agent. We have also lost a cheap method to purify our drinking water and a valuable mulch.

USES OF KUDZU

By ISABELL SHIPARD

Isabell Shipard.

Kudzu of the Leguminose family, has interested me for some time since studying the interesting Catalogue of Plants in the Appendix of Permaculture One. The Appendix lists a wide range of plants that could be useful in a permaculture garden, but it is often difficult to get hold of seeds, plants or trees, . Then a fellow Permaculture member lent me the *Book of Kudzu, a culinary and healing guide* by W. Shurtleff and A. Aoyagi, published by Autumn Press.

Kudzu is native to Japan and China, it has been found to thrive without fertilizer, pesticides, or irrigation. It has long been used for erosion control, and as a soil improver because of plentiful nitrogen fixing bacteria. It spreads on the ground, very rapidly making a thick mat, which has been used for green manure, and also as livestock fodder. This perennial can also be cut for hay (which is considered equal to lucerne and clover), or fed to cows to increase milk production. It contains 18% protein and is a good source of calcium and vitamin A & D, and C. Research in America has revealed that crops grown on land after kudzu have yielded 700% higher yield than without a previous kudzu cover crop. The beautiful wisteria like flower provides a good source of honey.

The tender young shoots and pods make a delicious vegetable, served as very young leaves in salad, or stir fry dishes, soups, etc. But it is the large root (which can grow over 2 metres long and weigh over 200 kg) which has been used in Japan since Ancient times as an ingredient in fine cuisine. The root is dug up, cut up with chain saws and diced, processed and dried, and esteemed by many as being the finest quality cooking starch and natural herbal medicine. It is used like arrowroot as a thickener for sauces, soups, makes a delicious crispy coating for deep fries: like a gelatine for setting, giving a very delicate texture and flavour regarded as an energy source. Also, the roots are considered very yang (alkaline) in nature and are found to be particularly effective in treating yin disorders. The root is also used in herbal teas, and coffee. Medically the root has been used for digestive disorders, colds, fevers, flu, diarrhoea, hangovers, for external sores, skin rashes, also headaches, sinus, thirst, anaemia, internal bleeding, and many other infirmities.

Warm and moist climates give ideal conditions for kudzu growth, and for this reason it has been called 'a mile a minute' or 'a foot a night' vine. If not used for its many advantages, it could well become a pest. But for those who see its resources and potential, it has been used to great advantage to improve soil, restore worn out land, erosion control, animal forage, human food, and mulching and green manure. In Latin America some farmers grow Kudzu among their coffee shrubs, apparently the vines do not climb the shrubs. It has also been used to reclaim semi-desert or land overgrown with weeds. It would be worth experimenting with it to choke out groundsel, and then with its nitrogen fixing properties the soil should be in good condition for crops and trees. Literature says that if the Kudzu is trellissed, there is then no danger of spreading via roots from vine nodes. Although it is perennial, it will die down in cold winters, and put on new growth in spring. It is

The Kudzu vine.

nematode prone. Kudzu has been found excellent for erosion control for embankments, and has been used to strengthen earth filled dams, reinforcing the infrastructure, while vines and leaves prevent soil washing on top, and vines will not tear at times of heavy flooding. Kudzu uses are endless: made into fine fibre (a kudzu kimono being the most highly prized and expensive), also used for paper making, fishing lines, basket weaving, building materials, stuffing cushions, making beds and chairs; and when burnt the fibre acts as a mosquito repellent. I don't think we will be experimenting with kimonos, but it has many possibilities for use on our farm.

A GRAIN PLOT

A Fukuoka Grain Plot Gone Wrong
Or How To Prepare An Orchard Site Without Really Trying

By KIM CHRISTIE

I'm not sure whether this is a description of a Fukuoka grain plot, a method of soil improvement or a poultry forage system, but it may be of interest.

Two years ago I decided to try to establish a Fukuoka type grain plot on about 1/8 of an acre, which was hard packed soil growing very yellow and stunted bladey grass, couch and kikuyu.

MID-WINTER

In June the plot was ripped once with a chisel plough – I could only get the tynes down about 4 inches – and the following were broadcast by hand:

    Lime at the rate of 2 1/2 tones/ha
    Dynamic Lifter – 3 bags for the site
    Wheat at the rate of 80 kg/ha
    New Zealand white clover at 8 kg/ha (innoculated)

The rotary hoe was then run once over the lot on a shallow setting.

A good stand of wheat grew which I started to harvest, but I did not have time to finish harvesting or threshing, so chooks and ducks were given access to the plot. The poultry harvested the seed for themselves and at the same time trampled the straw down. During a period of about 2.5 months the poultry required no additional feeding and grew fat.

SPRING

When all the grain and most of the straw was down I broadcast grain sorghum by hand on a wet day at about 60 kg/ha and by early summer there was a good stand of clover and sorghum in which I tethered the goats, having decided that this was of more use than the grain would be. This kept 4 goats going for most of 3 months as the sorghum kept re-growing.

NEXT WINTER

In June I again broadcast lime and wheat at the same rates, the performance with the poultry was repeated.

At this stage, I decided that the land would be of more use as an orchard site, so I allowed it to grow through the summer. There was a dense stand of clover up to 30 cm high right through the hot weather and this was mown twice to add organic matter to the soil. A very high setting was used and mowing was timed to catch any weeds just before they set seed. The poultry also fed continuously on the clover during this period.

The area now is a very much darker green than the surrounding land with no sign of bladey grass or couch, but a small amount of kikuyu with dense clover

and various other grasses. The soil is black instead of bright orange/red and the chisel plough will bite full depth. The surrounding area is still mainly poor bladey grass and couch and the ground so hard as to be practically unploughable.

This method may be of use to anyone wanting to improve some land without much work or cost, while at the same time gaining some production in the form of stock and poultry feed.

COMMENT

Kim's farm is situated about 6 km west of Nambour (S.E. Qld) at an elevation of approx. 400m. The farm receives some frosts (to about -5 degrees C) most winters. The soil is of volcanic origin, but due to the previous owner's poor management had been badly eroded.